Replacing something that's genuine, aged and valuable with something similar but ultimately a poor copy of what is already available is not without it's perils.

And if you try to force Google's hand where it is ranking sites, Google DOES have a big surprise in store for you!

Surprise

The 'rules' clearly state (in 2016), that if you try to manipulate your rankings through ways Google disapproves of – they will penalise your site – for a long time – and Google disapproves of a LOT. Making a claim for a top spot IN A COMPETITIVE industry without quality links and relevant content over a PERIOD OF SUFFICIENT TIME, in a vertical with relatively stable rankings, raises a red flag to Google. I've seen sites rise and rise and rise and when they get to the top, they get slapped back 40 places. Sometimes immediately – sometimes a few months later.

If you are at the top of the results, you can bet Google will take a closer look at your site. That might be deeper algorithmic analysis of your site, or even a manual review…. sometimes I actually worry about all of a sudden appearing near the top of results. Sometimes, it is shortly followed by a big drop, if the methods used were a little ropey.

If I bag a top ten ranking, I don't usually push for number 1 in Google anymore – not without a strategy based entirely on making things better – low-quality link building, for instance, is just not a long term plan I want to invest my energies in any more. I normally concentrate on other keywords when I get into the top set of results, and on building domain trust, and usually only focus on the main term if I have a solid gold linking opportunity on a site with mega trust. Why Is A Top Ranking In Google So Valuable?

Money!

A number 1 ranking in Google: attracts the lion share of visitor clicks and gets

a lot more clicks than no2 position, and vastly more than the other 8 listings in the SERP

Of course, that's assuming your search engine results page (SERP) snippet is as 'clickable' and 'relevant' as the competing pages' snippets for that search query.

Organic listings as a whole get more (perhaps double) the clicks a sponsored ad listing attracts according to musings in the SEO industry at the moment but it suits Google to balance that out in the future (because Google makes more money from advertising). I would be more specific with the numbers, but I don't trust most stats out there these days about such things.

Everyone wants to know:

How to get to number 1 on Google?

…but the truth is Google changes what is number 1 in SERPs pretty often.

Typically there a few obvious ways to get to number 1:

Or number 1 ranking in Google natural listings is most valuable, because you do not pay for the clicks. Free traffic from Google is the holy grail. Websites with a lot of organic number ones get a lot of free traffic from Google.

In competitive niches, you will need to pay Google to be number 1 using Google Adwords, and this will continue to be the case as Google becomes more an more, a local search engine (IMO). Google Adwords is typically the fast way to get to number one for valuable and competitive keywords and keyphrases.

What Do You What To Rank Top For? Some companies want to rank for different things and certain links and strategies achieve different results. That's the end of link building for beginners month on the Hobo site – hope you found it useful.

I'd thought I'd close mentioning you really should have a specific goal every time you start link building.

What do you want to rank no1 in Google for anyways?

your company name or brand? Not only do you want to rank for it, you want to control every mention of your company name in Google, easily done by social media participation and using the authority of other sites to rank for your brand. Very easy to achieve with just on page optimisation and a few incoming low-quality links (sometimes, not even).

Your service, in your area? – again, fairly easy. Done with on-page optimisation (geographic mentions in the title and in the text for instance), and generally speaking some low-quality links

Your service in your country? – slightly more difficult than above, but can be handled with plenty of low-quality links from even low-quality, unrelated sites in some cases

Your service? – difficult depending on the niche – you're going to need some decent links or at least the same amount of crap links your competitors have. Crap anchor text links outweigh unfocused poor anchor text links from even relatively authority sites.

Your products? – generally speaking, very difficult, especially if your products can be bought in a 1000 other places. You're going to UNIQUE CONTENT, need links that pass PageRank, anchor text and trust ie ranking ability. You're going to need a few trusted sites to link to you to rank all those products. The more pages on your site, the more Pagerank you will need. To get pr, you need incoming links.

A weight of crap links built over time can beat even a relatively trusted site in Google in 2009 – still. However, it's these links Google have a lot of brainy people working on attempting to nullify, so why swim against the tide? ESPECIALLY considering just a few links from one site can transfer instant ranking ability and trust to a new site, or one link from one PR5 page can transfer enough PageRank to heat up an entire

200 page website with no other links. Finding those sites can be a full time occupation though.

Deciding what you want to rank for and how you want to do it are at the core of link building strategy.

Search Engine Optimization

Search engine optimization is the process of making pages 'as relevant as they can be" for search engines to believe they are valuable enough to be considered for top rankings for as many key-phrases as possible in organic or natural listings.

Nobody knows for sure how to get number 1 on Google, not exactly anyway, but getting to number one in Google is largely down the reputation of your website and how relevant pages are to keyword queries. This of course needs to be manipulated to get the best out of a site, and that's where companies like Hobo come in.

Typically you get to number 1 by having a good online reputation. Big brands have good reputations. Big brands rank at the top of Google, too. Your reputation is increased by the number of quality web pages that link anywhere to your site. Typically relevant pages with the most, and sometimes the best, links rank at the top of Google natural / organic listings.

Instead of focusing on number 1 in Google, your focus should be to appear in as many Google properties as possible, to give your business as much opportunity as possible to appear for as many searches as possible that are relevant to your business. FOr instance, we are a SEO company, so ranking for 'hot to get to number 1 on search engines like google, Yahoo or Bing might well be valuable to us.

How hard is it to get to number 1 Google? Ultimately this depends on the competition for the keyword or keyphrase and the reputation of your website. New websites typically find good rankings hard to come by in Google in competitive verticals.

Placing number 1 on Google and getting no traffic? You must be number 1 for a keyword that is not widely searched for.

You searched for 'how to get to number 1 on Google' and I hope this Book has shed some light on this – at least you know you should be asking:

how to get to no1 in Google organic listings?

how to get to no1 on sponsored listings in Google?

how to get your site number 1 on Google news?

get number 1 listing on Google maps results?

how to get number 1 on Google video results?

how to get to number 1 in Google image search?

how to get number 1 in Google shopping comparison?

how to get website 1 in Google blog search?

how can i get a site to number in Google updates?

how to get number 1 on Google local business (now Google Places)?

You can get to number 1 in Google for free if you know what you are doing, and if you don't, you can pay Google Adwords or find a search engine optimiser like Hobo who can consult with you to help you rank no1 in Google, Bing & Yahoo search engines' organic listings.

Beware SEO companies who promise no1 ranking guaranteed. No one can guarantee no 1 rankings in Google.

How Much Traffic Will I Get From A Number 1 In Google?

I'm a Link Builder, Jim, Not A F*&^ing Soothsayer!
I was cornered and asked this yesterday :
How much traffic can you guarantee I will get from a number 1 spot in Google natural (unpaid) listings?
and this seemed like the most accurate answer I could give…..
….more traffic than you would get if you were in number 2 position
Most keyword data tools are inaccurate. Without actually being number 1 for a term, it's impossible to say for sure.
Google Adwords of course is very useful

(the best keyword research tool?) but your paying a pretty penny for all that intel and often, the client has no data to share.

If you're a SEO with a few number 1 terms, you can look see what traffic you get and compare it with any keyword data tool for that key phrase, and then make some assumptions from any normalised data. You can also mine for opportunities in your niche comparing keyword volumes with keyword competition etc etc. – you can even look at Alexa (!) and Compete and a few other places.

I use SEMRUSH and Google keyword tools to give myself an idea of the most popular keyterms, but then again, I am a

link builder, not a keyword monkey, or a soothsayer.
I only need to do the easy bit – change rankings for as many keyword terms as possible – usually by building domain authority.
I hate wasting too much time on something when it's little better than a surmising that's been calculated. I'd rather spend that time thinking about getting real links from real sites (where possible). 6 months ago Google told me a keyword sent 20,000 visitors a month – today it is 18000 – local, it is 9000 searches. I am no2 for that term plural and single word) after wikipedia now in google.com, and getting 200 visitors a month from both terms.

That's a pretty big discrepancy even between the local numbers. I got 10 visitors on that term today. 10×30 is 300. Even if I triple that for the number one spot, that's still a bit difference!

Then again, I get 11,000 visitors a month to a term/keyword combo that doesn't even register in any keyword tool lol!

I certainly never trust the actual volumes I'm told about by these keyword tools. Often I'll turn to Google Analytics to spot more likely/ achievable opportunities in traffic.

But as an SEO, I'm supposed to know about keyword research, linkbuilding, server responses, Pagerank, relevance factors, ranking benefits ET AL.

When I am speaking to potential clients, I'm usually in, just after some swanky salesman from another SEO company, has hit the prospect with GUARANTEES and FIGURES I think to myself I'm not even bothering trying to compete with this because most claims like this in SEO are nonsense – a best guess at the very best.

And I'm not saying keyword research isn't important. It is a fundamental part of any campaign – perhaps the most important part. It's just not my favourite bit of SEO (getting real links to relevant pages is), and I hate making any kind of predictions in any market I don't have years of experience in – and there's a lot of markets out there.

I pointed out to the prospect if whether I win this contract or not is based on whether or not I can guarantee anything about traffic levels, I'll bow out of the negotiations. Whatever I come up with is a best guess.

We did win the business….. proof you don't always need to talk blocks to get on in SEO or online marketing.

Number 1 Ranking In Google + Bing + Yahoo

Here's a secret some SEO companies might not want you to know about. The number 1 reason a lot of sites get number 1 places in Google listings are….

…they are good enough. It really is that simple.

This simple:

Basic good practice on page SEO guidelines.

Sprinkle a few high-quality anchor text savvy links at the site while avoiding like the plague any "get links fast" schemes

ENSURE on-site navigation is google friendly, and coherent.

Ensure the site is easy to update

...then, round of with some killer, original content that is of at least similar quality to other sites and pages competing for the desired terms and you know what, after rinsing and repeating ad nauseum, you'll probably rank in Google for desired terms.

I am often bemused when I visit Webmasterworld and see poor Webmasters who've been banned by

Google (which i suspect happens a lot less than people think) or have completely lost previously high Google rankings. I automatically think to myself, if I visit these sites, will I be disappointed? Do they have the content or is their position "link generated"? If they do have the content, have they screwed up the simple stuff? Have they made a big mistake, somewhere critical? Google doesn't owe anybody good rankings. Just because you've been number 1 for years doesn't mean the position is unmovable, either.

And anyway, is there such a thing as "Number 1 in Google" these days? Ever flux and constant updating ensures a freshness to the top ten, so in reality no

SEO company can guarantee any company prolonged number (insert number or page) listings – although many do.

Every position in Google is up for grabs. Instead of buying links, mass link-bombing of keyterms, mass registration of fake domains or trying to "game" Google, why not just add good content to your site and make your site "better"? Don't worry – with practice, it gets easier. In time, rankings come – but only with good content.

Sure, some sites bend the rules and get good rankings. They may even keep these rankings for some time. But in the end, they are generally hit by changes in the way Google deals with things.

That's what we help customers with. We consult with them to try and make sites better. For Google and visitors. It's a long term strategy, takes a lot of hours, but surely it's the only sustainable method for Google success for most companies.

And you know what? It works.

We don't know how Google works. No SEO company does or can know (and if they do, believe me, they are too busy making money for themselves to help you). We are successful because we know the kind of sites that Google likes (this is common knowledge), and we help clients to try and develop these.

Sometimes it's knowing what not to do rather than what to do that gets listed at the top of Google.

If you want us to do your SEO, you'll need to understand this. If you have a one page website, basically an advert for your company, we're not willing to spam the search engines for you, as this isn't common sense sustainable business for you or us.

There's too many customers out there who are willing to spend the time required making their website quality for visitors and Google. We're too busy helping them.

If you're one of these clients who want to make their website better, contact us. If not….

How Many Clicks Does A No1 Ranking In Google Get Compared To No2, 3, 4 & 5? A LOT more, that's the only thing that can be bet upon. It really does depend on a multitude of variations, from what Google displays around your listing, to the nature of the query itself.

Google Webmaster Tools now shows click though rate and position in SERPs – so you can work this out for your own site. Not that it's accurate – but what else do you have?

I picked a term I know I have had the top 5 positions at various times, and it's interesting to see the clickthrough rate on particular keyword searches…. and how many clicks the top position in

Google gets compared to the number 3 position, no4 and no5.

Position 1   58 46 79%
Position 2   91 46 51%
Position 3   210   73 35%
Position 4   260   46 18%
Position 5   110   12 11%

Obviously, this is just one example – it will take a while to look into the new data and look at an average – but it shows a number 1 getting nearly 30% more of the clicks than a no2 ranking. You might find some useful nuggets of information at Google Webmaster Tools for your own site…..

Of course, click through rate can be skewed by any number of factors – the nature of the query or how compelling your call to actions are in your title and your meta description, to name just a couple.

This info might prove interesting once aggregated.

Get Top Ten Rankings In Google With Simple SEO

Simple SEO is just that. Simple. You can get top ten rankings in the SERPs in many industries just by following some very basic (on the whole, onsite) SEO tips.

It's worth pointing out that you *typically* have three chances to tell Google what a page is about, and how important the page is.

On Page – The actual text content of the page

On-Site – In internal links to the target page

Off-Site – In links to the target page from other websites

OK – You've got your site….. it's got the usual stuff – home page, contact page, about us, map, products – but you have a blog! A blog lets you easily add pages. That's all you really need, although you can do this without a blog of course, but then you need to know a bit about website design.

There may be some evidence that the more you link to a page in your website navigation structure, the more important Google seems to think that page is, in relation to the rest of your site at least.

Pages that aren't linked to frequently may not have enough link equity to make it into Google's main SERPs.

Optimize 1 page for 1 keyword (multiple related key phrases)

Make sure you have a keyword rich page title, the words and keyphrases on the page and in the name of the actual file path if possible

Link to this page from within your site with the anchor text "keyword" a few times at least

Don't link out from that page with the exact anchor text "keyword"

Going forward, try and encourage other sites to link to this page with the anchor text "keyword" as opposed to your home page. This is called deep linking.

Of course, the more unique and better quality the information on your page, the easier it is to achieve this. Stay away from low-quality link sources.

Sometimes, I consider linking out (where relevant on this page) to other quality sites

Thinking: A well optimised page followed with a few incoming links from external sites will perform very well in Google, and is boosted when you tell Google "Hey – This page is important", by linking to it from other pages on your site. Not linking out to any other page (from the target page) with the exact term you are targeting tells Google as far as this page is concerned, it's the authority document on the matter (which is the aim of SEO). You would think.

Warning: This works well for small sites, with a few products. Using this strategy on a site with a lot of target pages will have mixed results, and you risk making the site look spammy.

Simple SEO might be all your website needs to get better rankings in Google. Always remember not all links are equal. Nothing helps an individual page more than on-topic links from reputable websites, but it's clear you don't need thousands of links to get top rankings in Google.

Make a relevant, well optimised page that is well linked to in your internal site structure, and back it up with a few anchor text rich links from external sites. This strategy helps leverage the overall authority of your domain to rank specific pages, ideal if you've not a lot of authority to begin with.

Tip to Remember – Give Google what it wants – Optimise your page, and always link generously to your important pages within your site navigation and content. BIGGER TIP – DO NOT OVERDO IT. Keep it simple.

Traffic Is Are Never Guaranteed
No1 Ranking in Google Lost 87.5 % Of Value In Last Month & It's Still No 1.
A lot of folk have been complaining about loss of traffic on especially long tail searches. My initial thought was to do with internal linking because that's

how I've traditionally 'optimised' for the long tail & increased SERPs competition etc – but a lot of different things could be at play.

There's an interesting discussion at WMW.

I've been digging about analytics to see if I could identify a particular reason for this (as I see it on a few sites I monitor) and there doesn't seem to be anything stand out and consistent in my analytics so I checked the source – Google SERPs. In one example I thought was interesting enough to share – I'm looking at a 4 keyword term I am number one for (which gets a bit of traffic) drop 87.50% in traffic in the last few weeks…. AND IT"S STILL NUMBER ONE.

LOL

What do I have to thank for this PARTICULAR SEARCH?

New 3 column SERPs layout making SPONSORED LINKS creep that further bit down the page and push organic listings further down

NEW SERPs layout encouraging clicks away from the organic "centre' listings – I mean, come on WTF is using that 'wonder wheel'?? Totally distracts the user clicking into Google UK only results Google Shopping Results and local Business Results & &^%$ing video results and images results, PROBABLY grabbing folks attention from that coveted(?) no1 slot

Hey lets not forget REAL TIME UPDATES & NEWS & Search Customization Updates – the list is extensive

Oh yeah, about that number 1 slot – barely above the fold tonight as I check…. still number one though for all it matters but when you're committed to building good solid sites for customers and aim to increase month on month traffic it's not nice to report back:

Oh 87.50% DROP on a main keyterm traffic – Google has f*&^%$£ you. We need videos, pictures and shopping feed to feed the Google monster these days

Now – of course this is on one keyword and it's a bit of a SENSATIONALIST TITLE I'm using, and everybody will have different reasons for drops in traffic –

and plenty are moaning about THAT over the last months. I just thought this was a single, granular example of how I lost a lot of traffic on a keyterm just because of UI changes.

It's not just what YOU do – it's what Google is doing with those SERPs too.

In some cases, loss of traffic IS probably to do with how Google is presenting all it's products to searchers too.

It's reasonable to assume as Google refines it's products, they are going to seep more and more into longer tail searches, stealing clicks when previously, you would have got them.

Whats the point of being number 1 for a term if just below or above that Google is presenting eye-catching distractions via Google Video and Google images and Google News and Google Local Business Listings?

It's probably never been more important to make sure you are taking advantage of ALL Google channels these days because Google is – and it's playing about with where they appear on the page.

On a seperate note I do subscribe to a lot of the long tail traffic drop theory out there at the moment too – and would probably think Google is getting better at crawling deeper and faster too, and identifying better links, which is could well be the reason if you are experiencing traffic drops.

Why Do Google Rankings Change All The Time?

A big misunderstanding of Google and search engines like Yahoo & MSN is to view them as "one big super-computer." In fact, they are tens of thousands of machines, located in different "data-centers" (DCs) all over the world.

And they do not get updated all at the same! Instead, changes are rolled out slowly, a few data-centers at a time, and a few machines per datacenter at a time. As a result of DNS-based load-sharing, the "Google" you connect to right now is not the same "Google" you connected to five minutes ago — It is a different machine at a different IP address so different set of results.

So, you are simply seeing results on different Google machines, depending on when you connect (and where you connect from).

If you see your brand-new site appearing and disappearing, but ranking well with increasing frequency, that is potentially good news.

On the other hand, if you see your well-ranked site dropped with increasing frequency, then that is bad news. It is however possible that you're connecting to only partially-updated servers (computers), and your data isn't loaded yet. It doesn't make sense to panic until your site disappears completely, because it might drop, or it might pop back — You just can't tell.

This is why no SEO company in Scotland can promise you No.1 in Google. From minute to minute, even Google engineers don't know who will be top for a specific search term on a specific computer / datacentre.

We aim to build good quality sites with quality incoming links to ensure at least your site remains bobbing about on Page 1 of the results.

Does Google Play Loaded Dice With Your Rankings?

The Google Algorithm

What Would You Do If It Did?

Recently I found an exact match domain for a little project, with 4 domain extensions. I popped a little bit of text on each so as to be 'unique'. A one page holding site for each. I left for Google to discover.

A few weeks later all 4 exact match domains are in Google(.co.uk) and ranking for their term in the same vertical (with another 168,000 results).

.com 5
.net 12
.org 21
.co.uk 23

Your rankings of course, will ultimately be determined by your content and incoming links, and the rankings will fluctuate, but it struck me as slightly interesting to see the difference in ranking between the sites, as I have often wondered where randomness factors into Google – if it does.

In the test sites, the titles are the same, the keyword is mentioned the same amount of times etc etc... theres only 50 words on each page max. There really is not much different between the pages – at all – apart from the domain extension. If you have a .com in this case, you are laughing – immediately in a top 5 position. But if you choose a .co.uk, you start from the 3rd page? Dead in the water. At least, your starting from a different point.

Perhaps it's to do with the domain extension, but perhpas it is an indication of how Google works at a granular level – the discovery phase – perhaps at this level, your positions are assigned randomly based on a particular set of principles (which we will never know). Perhaps this randomness is prevalent in Google inner workings and is what protects it from us ever finding out exactly how any particular element works, and even employees knowing it all, or being able to 'promote' – Matt Cutts did say on his blog:. someone walked up to me and pretended like he wanted to bribe me: $500,000 for a 1st place ranking. I turned him down, because no one can guarantee a #1 ranking â€" not even me. I've REALLY tried to isolate some fairly simple elements use in the past – some I thought MUST give me a definitive answer but alas they did not.

If this was the case it means trying to actually figure out how Google works is a non-starter – it would mean there was no sweet spots, anywhere. Perhaps it's different for all sites. For all elements. Join that together with some ranking elements that are turned OFF, or tweaked, personalisation, geolocation etc etc and you have something that can't be gamed. Well, too much. Perhaps this randomness is more diluted for the top sites, than the churn they sit on (everything after page 2 or 3)?
What would you do?
It's actually very easy to get good returns from Google unpaid listings if you give it what it wants.

You can stack the odds in your favour by adding lots of content and getting credible links to your site. That's what SEO is for me. Look at what the competition is winning with and try and figure out how to 1. compete and 2. beat them. Usually that means copying them to a point, and then trying to do something better at some point when inspiration hits.

Google has a lot of spammy verticals it seems to want to let you spam your way into them as long as it's a good relevant site which is at least as good as the current competition that have already. Of course, some verticals seem more protected than others, but that could just be the level of competition, or an 'age' thing.

It seems as if Google purposely uses brands to clean up verticals with lower quality competition. If I see a vertical with a lot of big powerful brands as the top ten I think 'hello' – here's a vertical Google needs some help with. Brands (well, specifically internal pages, like a bbc Book for instance, are good, but they don't beat focused anchor text linkbuilding on their own). Or even a great exact match domain that's been in a low-quality linkbuilding campaign.

To get the most from any element, you probably need to be a player – an online entity – a site with trust in any type of competitive vertical.

Which means getting BETTER, or more trusted, more credible links than the competition has, if your page is RELEVANT.

Google is clearly going to be using signals from brands for a long time to come. Links from online brands will make your website a brand until it finds another way of finding trusted sites.

I see a 'brand' as a real site, with some real links to it (or fake real links). This is probably why the SEO companies who put links in their client websites rank at the top of the SERPs. I don't ask any SEO clients for links, but I ask folk we've made websites for, for the odd link.

As soon as Google can access your pages, with simple navigation, with original content, with a good title – it really is about getting links from real sites.

This is all pure theory – just a mind wander if you are into SEO geekery. Don't go changing your domain name or anything silly. maybe this is all just a cae of – it looks like that – sometimes.

The point of my Book I think is to point out even though you don't have all the answers, getting quality links is and will be the most important thing you can do to get better rankings from google. If you want more traffic from Google meantime – add content. Lots of it.

How To Check Your Rankings In Search Engines

The most reliable mac/pc tools I've found are Rank Tracker & Advanced Web Ranking – both reviewed on my SEO tools review page here. Both tools are pretty cheap and they do what they say on the tin. Both have 'search engine friendly' rank checking modes.....Rank Tracker is simpler to use in my opinion. But if you are going to use tools like these to check your rankings, then you'll most definitely need a tool like HIDEMYASS to protect your privacy while you check your rankings – as most search engines don't really like automated bots like these scraping their content (the irony). Eventually, your IP will trip the limit and search engines will block that IP with Captchas. Which is when HIDEMYASS changes your IP. Where possible, it's always best to treat the search engines nice, anyway….

The ideal setup is probably to have a 'workhorse' computer with AWR or Rank Checker scheduled to check rankings at the same time every day – with HIDEMYASS installed and set to randomly change your IP location every few minutes between servers that are geographically local to you.

I save my rank reports to free web storage website DROP BOX so all my reports are accessible from all my machines.

One machine to run the reports. Many machines to view them.

Don't get too fixated on all of your rankings – some terms and results pages are bound to jump all over the place – Google is designed that way! And don't just try to rank for just a few terms. It's much better to rank for lots of different – and related – terms than have a business based on one keyword at the top of what's bound to be a competitive term…. so that means just adding lots of related content and getting some links to it.

How To Check Google Rankings In Other Countries

The Ad Preview tool from Google also shows organic search results and how they look to users around the world.

You can see just how rankings differ from country to country and place to place – worth considering when researching keywords. Geolocation and personalisation really mixes Google rankings up too.
https://adwords.google.com/select/AdTargetingPreviewTool

I forgot all about this tool until I recently saw it mentioned by AlyssaS on Webmaster world I thought it might be useful for some. There are desktop programs and online tools that allow you to do this too, and track your rankings over time. It's worth noting Google does not like rank checker programs.

Check Google, Bing & Yahoo rankings in other countries using Rank Tracker, Rank Checker Ace and AWR (Advanced Web Rankings) all reviewed on this blog.
I've collected my favourite SEO tools here.
PS – Use a rotating proxy.

SEO Companies Who Guarantee Number 1 Rankings Are Lying To You

Matt Cutts of Google explains a bit why SEO companies are lying to you when they claim they can guarantee number one rankings in Google for competitive terms….
What Matt didn't touch on was diversity in search results Google aims for, or geolocation differences, personalisation

re-ranking, or what the competition is doing, or tweaks to the algorithm, etc etc etc…. but it's true no SEO company can guarantee no1 rankings in Google, so don't be duped.

Taking the analogy of a Horse Race, even if you know where the finish line is, even if you have spent ages examining the competition, and even if you have 'fixed' the race so that every other jockey is in on it……what if happens if your horse falls at the second last fence? What happens if it breaks a leg? What happens if it drops dead? What happens if another faster horse enters the race at the last minute from left-field? You're No2.

And if you're caught cheating – you won't even be allowed to race.

Google is like a horse race, with hundreds of potential participants. It's a horse race where;

You don't even know where the finish line is!

You don't get to see everything the competition is planning to win the race, what they did yesterday or what they will do tomorrow, whether that be address site issues, get those all important quality links or indeed hire a better search engine optimiser!

It's impossible to know how healthy your site is in comparison to the no2 site, assuming your site is no1.

If you are caught cheating, your not allowed to compete in future.
It's tiring to hear…
"sure I can guarantee a no1 listing, if you give me enough money"
Yeah, throw money at it.
Google Guidelines state;
"No one can guarantee a #1 ranking on Google"
For me this is the only absolute truth Google tells us. Matt Cutts of Google is on record as saying;
"someone walked up to me and pretended like he wanted to bribe me: $500,000 for a 1st place ranking. I turned him down, because no one can guarantee a #1 ranking, not even me."
Matt Cutts

Even the cleverest of the black hat SEO brigade can't capture very competitive key-terms for long. Google will eventually catch up – Google manually checks some of the most competitive 'money terms' when the algorithm misses it and if they don't pass a human review – the site will be penalised, along with every other site in that bad neighbourhood.

If you're in business, you're obviously not stupid, or if you are, you won't be in business very long.

A SEO can't ever guarantee you top results just in the same way as an advertising agency can't guarantee that next advert in the local press will get you sales. If you're paying good money to a

SEO, you should expect good positions, but that's it. To guarantee anything you need to know all the variables – even SEO who claim you can guarantee no1 listings on less that competitive terms are fooling themselves. Google can take these positions away from them as quickly as a better SEO with a better site with a better back-link profile. No1 in Google is not an absolute, it's a floating point.

Best SEO Company Results in Google UK

The 'best seo company' search engine results pages always make for an interesting review as many are using tactics they can't use on your site. These companies do not expect to rank for very long, so it's common practice to spam a shell site into the top of results to get new customers.

It Now On Amazon@.com

1. Get visitors to interact to show Google that people engage with your site

Take Action:
Present your website content in the form of numbered lists with pictures so visitors have to scroll down to see it all, and break up Books into two pages so visitors have to click to see the rest of the Book.
2. Send a press release so news sites will link to your blog posts

Tony sent out a press release for a blog post he wrote about what he learned from his son who has Down syndrome, and other websites picked up the press release and linked back to his blog, giving the post more credibility with Google.

Take Action:
A day or two after you write a blog post, use a free service like PRLog to send out a press release that includes a direct quote from the post and the URL.

3. Diversify your backlinks to protect your search ranking

Tony says that if websites had drawn backlinks from multiple sources, they would have been fine when Google stopped indexing blog networks, but some websites relied on blogs for all their backlinks so their search rankings fell.

Take Action:
Open an account with Majestic SEO to check where your backlinks come from, and try to get backlinks from additional sources like social media or wikis if most of your backlinks are from one site.
4. Automate blog comments so you can get more backlinks and higher traffic

Tony uses software to automatically post a large number of blog comments linking back to his sites, and this makes the sites appear more popular to search engines and increases traffic.

Take Action:
Sign up with BlogCommentDemon, enter keywords you want to target, and have the program find relevant blogs and automatically post comments on them linking back to your site.

5. Use social sharing to look important to search engines

Take Action:
Create an account with a social media service like Facebook, and either pay for

other people to share your site on Twitter, Facebook, and Google+, or promote their sites in exchange for them promoting yours.

6. Stop using Google Analytics so Google won't see all your SEO tricks

Take Action:
Stop using Google Analytics, which could tip Google off to your SEO strategy and cause them to demote your site if they don't like your tactics, and instead download an open-source analytics package like Piwik.

7. Put content on .edu sites so your backlinks will have instant authority

Take Action:

Search for .edu sites that use MediaWiki, find old wikis that students are no longer using, and edit them by adding links to your sites.

8. Control suggested search terms so you can protect your site's reputation

Take Action:
Type your site's name into Google, and if it suggests any negative terms, pay workers on ShortTask or Amazon Mechanical Turk to search for your site with positive terms.

. If you're involved with SEO, you know that Google (and the other search engines) uses a complex algorithm to determine how a page ranks in the search engine results pages (SERPs). No one but Google knows exactly what's in the algorithm's "secret sauce" (it's rumored that there are more than 200 factors it considers), but SEO experts have inklings about what makes a page rank higher than others.

In the good ol' days (like back in 1998 when I got started in digital marketing), ranking high on the search engines was as easy as putting a bunch of keywords towards the top of your page and you'd suddenly find yourself in the #1 spot on the search engines for a highly competitive keyword. (If only it were that easy today!)

Now it's much more complicated when it comes to ranking on the first page of Google, and there are literally hundreds of variables and factors that affect how a page ranks on the search engines. Search engines look for "signals" on a web page to determine the worth and value of the page and determines how it should rank – looking at things like page speed, keywords, mobile-friendliness, words that are in bold, the number of backlinks, internal anchor text links, etc.

How Can You Rank Higher in Google?

Based on trial and error (and with some confirmation from Google every now and then) there are a few things that SEO professionals know can help pages rank higher. In no particular order, one of those ranking signals is quality content. Content that contains answers to and information about what people are searching for is one of the best ways to rank higher in the search engines.

Google also likes long-form content (around 1,000-1,500 words per page.) This means putting a quickie blog post up usually won't help your rankings. Put some thought and research behind every word you put on a web page. Find out what people are searching for and provide content that delivers those answers, and that will go a long way towards ranking higher for keywords you're targeting.

Another thing that can help you rank higher in Google's search results are backlinks. We're not talking about shady paid-for backlinks or directories whose sole purpose is to put up a bunch of random links to websites that offer no real value in and of itself (these things will only hurt you), but quality links from highly authoritative sites that link back to your website.

It takes work to get these quality backlinks, but it can be done and it's totally worth it. Link building MUST be part of your SEO strategy if you want to remain competitive. So if you don't have a solid content strategy in place, get to it!

Enter Google RankBrain

Now we come to one of Google's latest ranking factors: RankBrain. Google's new machine learning system is now a big component on how a site ranks. Machine learning may sound like something you'd see in a "Star Trek" episode, but machine learning is very real. Machine learning is when a computer teaches itself how to do something – and it basically gets smarter and smarter as it "learns." RankBrain is Google's artificial intelligence system that is now being used to process and compile search results.

RankBrain is a relatively new factor in Google's search engine algorithm. Google began rolling out this smart system slowly in late 2015, but Google has recently stated that RankBrain is now fully rolled out – a "living and breathing" factor in ranking higher on Google.

Google says that its new RankBrain machine learning system is now the third most important factor to ranking high on Google, and that RankBrain processes every search query. Think of this technology as a search query refinement tool – something that is learning the intent of what individuals are searching for so the results Google displays are the best ones based on the keywords used by each searcher.

Here's an example about how RankBrain might affect rankings: If RankBrain sees someone searching for "best pizza restaurant in chicago," it might determine that "best pizza in chicago" is more frequently searched for, so it may show those search results instead. What RankBrain "learns" about how people search can influence what results appear and in what order.

I won't kid you. RankBrain is still mysterious – but it's real. Only Google knows the full details and my guess is that even Google doesn't have it all figured out just yet.

Should SEO Professionals Be Concerned About the Power of RankBrain?

SEO has always been an evolving strategy for ranking high in the search engines. One day posting Books on Book syndication sites can help rank you higher. The next day you could lose rankings because of that Book. Go figure.

My guess is that SEO will always evolve making it even more important that we pay attention to what ranking signals the search engines are looking for, provide those signals in an ethical way and, most importantly, give a site visitor the information and content they're looking for. After all, content and information is what the Internet's all about; right?

There's no magic pill when it comes to SEO. The industry is morphing and always will. What we can do is continue to focus on putting the best content on our website and promoting that content through a variety of sources -- be it through paid advertising, backlinks, social media, newsletters and more.

My guess is that RankBrain will just make searches better for the end user, which has always been Google's #1 goal.

Start Now!!!

How many Backlinks do I need to rank on the first page of Google?

How many Backlinks do I need to rank on the first page of Google?

How many Backlinks do I need to rank on the first page of Google?

Update: You can now read the second part of my backlink experiment after 6 months. What works for Google and what doesn't?
Don't forget to subscribe to my newsletter (look for the op in box at the top of my blog or in the sidebar) to get updates about my experiment.

How many times have you asked yourself this question? Maybe 10, 50 or more than 100 times?

Well, I have good news for you. You don't need thousands or even hundreds of backlinks.

It turns out you don't need that much, and that's good because Google has changed its algorithm from quantity, to quantity and quality. But quality is more important now.

In fact, you can make it to the first page of Google with just a few backlinks and I bet you can count them with the fingers from1 your hands and feet. Do you want me to prove it?

Let's make an experiment!

After one year of creating Stream SEO, I've built around 4,100 backlinks (verified with Majestic SEO) and I have a few Books positioned in the first page of Google already. Some of them are even on the first 3 results shown, and those are bringing me a good chunk of traffic now.

But guess what?

I haven't build a lot of backlinks to those Books. Some of them have 2-3 backlinks. The best ones have 10-15 backlinks only. And I'm not the only one saying this. There's a recent Book at SEOMOZ explaining how a guy achieved this with

13 backlinks. All of them with different anchor text.

So, let's make an experiment again. I'll try to rank one of my recent Books on the first page of Google and see how it works. I might need some help from you here, but I'll do my best on my own and show you all the results.

I'm choosing a pair of Books I wrote back on December, to make sure they're well indexed right now. Those are my Createspace 6 Review and my Createspace vs WordPress Book.

Have a look at the keyword research analysis at Market Samurai. I'm using phrase match type and here's the number of visitors I could receive on a monthly basis while being at the first result on Google search results.

how many backlinks do you need traffic

So basically, that is 546 visits per month on the Createspace Review keyword, and 302 visits if I rank first on the Createspace vs WordPress keyword.

Right now, I'm already ranking with both of them on the first 100 results, but not on the first 10 results (first page). Here are my rank tracker results so I can be totally transparent with you:

how many backlinks do you need rank search engines

The first column shows Google's search engine rankings and then comes Bing and Yahoo. It usually takes a lot more to rank on Bing and Yahoo (a few months) while it takes me a pair of weeks to get into the first 100 results on Google.

So right now I'm in number 18 and 21, and as you can see, I've moved up and down on both keywords recently (red means going down and green means going up!).

Now this test wouldn't be interesting it the keywords were super easy to beat, or extremely difficult. Both keywords have some competitors. Some of them easy to beat and some of them not that much. Here's how it looks.

Createspace Review:
how many backlinks do you need competition analysis

As you can see, there are a lof of reviews out there, and one of them comes from a Youtube video. Also, results #7 and 8 are directly from the Createspace official page, but those are from one of their apps, not directly form their web hosting service.

Createspace vs WordPress:

how many backlinks do you need competition

Now this one looks more interesting. The results look easier to beat, but there's a powerful result coming from the Forbes magazine website. Aside from that, the first result comes from a page with exact match domain (who said Google's EMD update took all of them away?) and result #9 comes from Createspace forum.

My plan – How many backlinks do I need?
I'm going to start testing this week, from February 10 (2013). I'll try to build no more than 1-3 backlinks per week

(quality backlinks, no SPAM or spun content) and then track everything back with Market Samurai's rank tracker as I've shown you before.

Then I'll quickly update this Book every week and let you know the results. I want you to notice everything in the process, so I'm including the pages where I'm getting the backlinks and even the anchor text.

And about anchor text, I'll be very careful and use different anchor text every time, or at least keep my main keyword no higher than 25% of the backlinks created.

I'll try different keywords like:

Review
Createspace 6 Review
Createspace vs WordPress
Createspace vs. WordPress
WordPress vs Createspace
WordPress alternatives
and know knows what else depending on the topic
Weekly log
Week 0: February 10
Backlinks created this week: 0
Position on Google (Createspace Review): 21
Position on Google (Createspace vs WordPress): 18

how many backlinks do you need rank search engines

This was the first week and the Books had 0 backlinks. However, if I remember correctly, I created a few (less than 5 backlinks) while commenting on commentluv enabled plugins. Both Books are on the 2nd or 3rd page of Google and they've been there for almost a month or so.

Week 1: February 17
Backlinks Created this week: 2
Position on Google (Createspace Review): 18
Position on Google (Createspace vs WordPress): 14

Position on Yahoo (Createspace vs WordPress): 5
How many backlinks week 1

This week I decided to create 2 backlinks from different domains: Hubpages and Squidoo.

Those aren't backlinks on cheap comments or Books just made for get a backlink with a low quality content. I wrote 2 different Books which I think will rank well on both sites both of them related to Createspace.

For my Createspace Review both backlinks have the same Anchor Text: "Review"

For my Createspace vs WordPress comparison, one of them has "Createspace vs WordPress Analysis" as anchor text and the other one is "Createspace vs WordPress Comparison".

As you can see, both of them got better rankings and are know located on position 14 and 18 on the second page of google.

Additionally, I'd like to pop out that my Createspace vs WordPress comparison appeared on the first page of Yahoo this week! How? I don't know. But I've seen that many times when i have a good

Book ranked on Google, it will rank high on Yahoo and Bing too. If I'm on the 2nd or 3rd page of Google, most probably I'm on the first page of Yahoo and/or Bing already.

Less competition, I guess...

Week 2: February 24
To be Honest this has been a pretty hectic week for me. I've been working until late at the office and had no time to build backlinks to my Books. As you can see, I barely posted 1 new Book here at Stream SEO too. However, here are the good news:

Backlinks Created this week: 0

Position on Google (Createspace Review): 16
Position on Google (Createspace vs WordPress): 11
Position on Bing (Createspace vs WordPress): 4
Position on Yahoo (Createspace vs WordPress): 6
How many backlinks week 2

The new surprise here is that even when I didn't make any extra backlinks this week, I was still able to scale 2-4 more places on Google search rankings.

That's really good.

And also my Book is now ranking on the 4st result of Bing, while Yahoo decided to drop my Book 1 position only. So basically, I'm on the first page of Yahoo and Bing, and 1 position away to appear on the first page of Google too! I'm happy.

But there's more.

See that little number at the right of the URL on the image above? It's showing backlinks (BLP).

That means I have 2 backlinks for my Createspace Review but 8 backlinks to my Createspace vs WordPress comparison???

So I fired the SEO competition module and added both URLs to get the source of those backlinks and i found something interesting. All these backlinks appearing on my rank tracker are backlinks created by commenting 1 month ago or so on January/December. They finally decided to appear here and they're giving me some juice already!

Who said good quality comments on other blogs didn't work?

Anyway, that's how this business works. Work now and reap the rewards later. Also, I'm curious to know what will happen when my 2 Books created from

Hubpages and Squidoo are added to the algorithm. Those, and a pair of extra backlinks I'm creating the next week on 2.0 websites (each one with it's own original Book).

Stay tuned! I'm so close to appear on the first page of the 3 mayor search engines, and then I'm fighting to get into the first 3 results.

Week 3: March 3
This week I've been creating backlinks from comments only. No more posts until I see if they're getting indexed or not. Even though it's difficult to know.

Most of the comments I've done have a backlink to those specific pages using commentluv and my comment strategy.

Backlinks Created this week: 5 (comments)
Position on Google (Createspace Review): 22 (dropped)
Position on Google (Createspace vs WordPress): 3
Position on Bing (Createspace vs WordPress): 4
Position on Yahoo (Createspace vs WordPress): 5
How many backlinks week 3

As you can see, my Createspace vs. WordPress Book now ranks 3 on Google and it's on 4 and 5 for Bing and Yahoo.

At this point, I'm getting around 20-30 visits per day on this Book, which means I'm getting almost 1 thousand visitors per month. If you watch my first impressions on the number of visits I researched with Market Samurai, you'll see that originally, I was expecting to get 300 more visitors when ranking on the first result, however I'm getting much more than that thanks to other long tail keywords which I'm not tracking yet.

In fact, just as I'm writing this update my keyword seems to be ranked on number

2! (Market Samurai's Rank Tracker updates each Thursday.

How many backlinks week 3-2

This is why I always recommend to look for long tail keywords and expect at least Exact Match results (don't expect broad results to appear ASAP).

How many backlinks week 3-1

My Createspace review has dropped a few places and this looks natural to me. I didn't build any backlinks for this Book yet because I got happy about my results. It dropped 6 places and I hope to get it back later. For now, I'll focus on the other Book as it's giving me results already.

Week 4: March 10
I went out to Cancun (yeah!) for business reasons I really wasn't able to work on my experiment that much. That said, I was able to create at least 3 backlinks from comments, especially targeted to the WordPress vs. Createspace Book. Here are the results for this week!

Backlinks Created this week: 3 (comments)
Position on Google (Createspace Review): 36 / 42 (read below)
Position on Google (Createspace vs WordPress): 2

Position on Bing (Createspace vs WordPress): 0 (dropped out from Bing)
Position on Yahoo (Createspace vs WordPress): 4
How many backlinks week 4-1

Let's start with the Createspace vs. WordPress Book.

In short, it's doing great. This week I'm receiving between 40-50 visits per day just for this Book, which is WAY HIGHER than the original expectations from Exact Match results. This means I'm getting around 1,300 visitors per month just for this Book (4 times bigger than expected) from different long tail keywords. I wonder how much traffic will I receive if I hit the number 1 spot.

Note aside, I've been dropped from Bing results this week. I don't know why, but still, Bing wasn't really giving me traffic. Google is my primary source and it's rocking.

How many backlinks week 4-3

But if you didn't notice, I now have 296 backlinks to this Book! How is this possible if I haven't build that much?

I fired up the SEO competition module and discovered that many backlinks are coming from a blog I don't even know. I think I left a comment one day and since this blog is using the "Top Commenters" plugin I might have appeared on the sidebar for every page, creating so many backlinks at the time and now they're indexed.

This is funny, because I'm not there anymore, and I wonder what will happen later when Google can't find those backlinks anymore. Time will tell.

Here are my results for my Createspace Review Book:

How many backlinks week 4-2

Ugh, looking bad.

So bad that even my Createspace comparison is ranking better than my review at this moment for the same keyword!

As you can notice, even though I build a few backlinks they're not indexed yet, so I still see only 2 backlinks on my Market Samurai Tracker. I guess I'll put more attention to this Book now that other one is doing good alone.

This is where I could use some help I'm not asking you to create backlinks or anything like that for me. But if you find this experiment interesting and would like to give me some feedback or simply want to share it with another person, please do so.

Share it, +1 it, like it, pin it or whatever.

I'd just like to make it easy and I'll report updates every week. I want to probe I'm correct (or maybe not) and show you how I positioned some of my Books on the first page of Google.

This can be done again, and it's all about creating quality backlinks. Forget about buying thousands of backlinks or using softwares and black-hat techniques. Those won't help you. And if they do, it won't last for that long. I promise. Shall we start?

I'm a full time Affiliate Marketing Expert and blogger in my free time. Follow me on Twitter or subscribe to my newsletter for more content.

I know some people make things harder then it needs to be. People tend to harp over (I cant get backlinks...) however, they tend up making mistakes long before trying to get backlinks they may not help their cause. 1: Niche research is very important. If you get this wrong, then you will fail, it is that simple. 2: keyword research can lead to disaster also if you do not target the right keywords in your niche. You need to target low/med comp keywords for your niche. Do not target keywords over that have metrics over 30. ie: if you target keywords that have metrics of 60, then your efforts will be just a waste of time, even if you target keywords with 30, you may still find some RESISTANCE,but you would be able to rank these keywords in your lifetime. target niche related keywords 30 or under. 3: bad/poor content; Over the past few years, google has been cleaning up the internet, bad links, spam, low quality content, etc. Your content needs to be shap, and provide high value to your visitors; If you can solve problems, then i can rest assured you will get backlinks as you would not have to build them, everyone else will be building links for you. There are other factors, however, I am out of time for today. here is a tip: build your house with solid stones, and your house will have will have far less risk of falling down. these basic skills are a must, they are not hard, however, seo is a time consuming process.

Hi Silva, Im a new blogger and As i came across with a lot of tutorials and advises of Pretending gurus online, I just sometimes came to the point that I became lost. Dont know exactly what to do and to follow. First was with launching a blog up to generating traffic. But As I came across here, I noticed something about backlinks, someone said that I should build a lot of that. But I will follow your advise here and see how this will work with mine. LooK, I failed with my first blog and my only goal is just to earn an income to sustain my daily needs as I dont have a job right now. Im thinking that this strategy will also worked with mine. thanks you for the great post any. Been following some of your post as well.

Comments
Woostah says
July 29, 2016 at 2:26 pm
Thanks Silva, I appreciate your efforts to show us how many back links to be first on google..

Comments
Jess Tiffany says
July 27, 2016 at 1:54 pm
Great information. I have recently been quoted in a few bigger sites for my company and was curious what impact it may make. Thanks for the insights.

Jess – President of the Marketing and Networking University

Comments

Sagar vyas says

July 11, 2016 at 10:27 am

Hello,

Got your email regarding "how many backlinks do I need". It was the nice Book. My question is if we need quality backlinks for boost ranking in keywords. But how we get quality back-links. In 2016 directory submissions is not good for backlinks.

Answer is only social media marketing for backlinks ??

How to get quality backlinks in 2016 ?

Thanks.

Comments
Tony says
July 12, 2016 at 10:02 am
Doing guest posts, PBNs, social media, building many sites, etc.
Directory submissions don't work anymore, but all the rest are still pretty active.

Comments
vinayashree says
July 10, 2016 at 2:55 am
Seems Innovative Idea Servando! Though there is no such critrea in order to determine number of backlinks for acheving top slots in search engine; still

relevency is the key of the success for ranking of keywords.

Comments
Victor Onokpasah says
July 3, 2016 at 12:50 pm
Hello,
Your Book is on point this is how easy one can actually create backlinks. Aside from squidoo and hubpage can you recommend others.

Comments
Utkarsh Mk says
June 26, 2016 at 11:47 pm
Nice Book This will help in my next event

Comments

Barry says
June 23, 2016 at 12:13 am
Interesting research. Thanks for sharing your findings!

Comments
Himanshu says
June 14, 2016 at 10:40 pm
Very Nice Book sir.
My question is that i have to use the anchor text to be the same with the keyword that is ranking high or that is searched more. like i want to rank website for android apps so i have to use the anchor Android Apps ?

Comments
Tony says

June 14, 2016 at 10:54 pm

You should use it but only on a percentage of your links or you will get penalized. You can use it like 10-30% of the time and the rest you can use long tail or related keywords.

Comments
Mahesurya says
June 10, 2016 at 4:25 am

Thanks. Nice to have the ideas all in one place!

Comments
DURP says
June 6, 2016 at 6:28 am

Great ideas. I need more information on getting more backlinks. And this is a great Book.

Comments
TechAndroids says
May 31, 2016 at 7:12 am
Hey there, really nice Book. I just started techandroid.com and i've been struggling to fetch some traffic. Can you tell me, how long will it take for my website to rank in search engines? It is very hard to get links as the site is relatively new.

Comments
Golam Maruf says
May 11, 2016 at 11:41 am

I love this Book, I always focus on my quality Book, Please give us some information about quality Book. thank you

Comments
Dani says
May 3, 2016 at 8:37 am
Very good Book, There is a list I have created with over 1000 high quality backlinks that you can implement if you so want. let me know if you need it.

Dani

Comments
Sumit thanai says
April 23, 2016 at 2:46 pm

You are master in this. thanks bro.

Comments
gina says
April 20, 2016 at 4:15 pm
I don't build links manually. I just focus on quality Book. Yes, I need time to gain backlinks. But it is free penalty…

Comments
Paula says
March 31, 2016 at 1:17 pm
Sir my website is real estate Qatar.my website category is classified how many back links need #1 rank in Qatar google search

Comments

Tony says
April 1, 2016 at 10:26 am
I have no idea. You can't really ask someone like that and expect a complete answer without saying much. Have you done keyword research?

Comments
Fernanda says
March 24, 2016 at 5:00 pm
Hello, I would like to know if this technique still works? I'm doing a PBN and creating backlinks in sites like hubpages. Thank you very much!

Comments
nobel says
February 29, 2016 at 11:10 am

really helpful for new learners who are trying to seo their sites..These backlink site lists are very effective… My alexa rank upgraded a lot trying these sites….

Comments
Garima says
February 11, 2016 at 2:57 am
GREAT BLOG.GOOD TO CHECK WHAT EXCTLY BACKLINKS ARE GETTING CREATED AND WHAT MORE PROGRESS WE NEED .. GOOD TO HAVE THIS BOOK HELPFUL IN MANAGING BACKLINKS. THANKS .

Comments
Johnson joseph says
January 19, 2016 at 10:43 am

Great Book Boss, sincerely this post has been configured to assist bloggers like me. I have few quality Books on my blog that should be ranking well but due to lack of sufficient back links they are not getting the desires ranks from Google. Considering backlinks from commenting on authority blogs like you'rs. How do you get back link back to your blog, since specific links to your Books are not included in the comments?..,
Secondly, should all my quality Books get relevant backlinks to boost ranking?

Comments
Tony says
January 19, 2016 at 11:06 am

Sometimes you can leave backlinks if it's definitely useful in the comment. Otherwise there are ways to get backlinks by talking with the owners and doing outreach/guest posts, etc.

Comments
vijay patel says
January 15, 2016 at 2:31 pm
nice information still work backlinks rank

Comments
brýle says
January 13, 2016 at 6:21 am
I think that its not about amount but quality and constant work..

Comments

TurnOnYourBrand says
January 3, 2016 at 9:55 pm
My website got ranked in the first page of Bing and Yahoo search engine at the very beginning stage. But still struggling to rank in google. A contradiction to thing which you have said "It usually takes a lot more to rank on Bing and Yahoo (a few months) while it takes me a pair of weeks to get into the first 100 results on Google."

Comments
Tony says
January 3, 2016 at 11:14 pm
Yeah, Sometimes it seems like Bing/Yahoo pick websites faster and sometimes it's the opposite.

Comments
Rushabh says
December 28, 2015 at 1:17 am
Nice information sir , actually i was confused about how many backlinks needed but finally i got right information, thank you.

Comments
Mesin cutting sticker says
December 25, 2015 at 9:44 am
Just reading this Book, and just know about hubpages.

Did write in a hubpages have a great impact to get a backlink?

How many Book should be write on a hubpages? I was write at least 2 Book there.

Comments
Alan says
December 10, 2015 at 6:41 pm

Comments
Jenine says
November 19, 2015 at 8:35 am
I'm familiar with the concept of backlinks, but how do you get them to your site? Do you have to comment on other blogs and leave a link back to your website or what? I really don't know how to do this.

Comments

Of says

December 5, 2015 at 7:53 am

Great Book – but this is exactly what comes to my mind – how to create back links ! Where to go and how to do it – hopefully he post something about that. Because knowing the importance of having the back links but not knowing how to build them is not so helpful.

Comments

Utpal Konwar says

October 17, 2015 at 1:50 am

I am create first time a blog…this is very nice helpful topic…thank you.

Comments

Santral Anons says
November 5, 2015 at 6:29 am
Thanks for the Book, it was nice…

Comments
sohag says
October 14, 2015 at 5:20 am
nice post . I have some back-link but Alexa show only 1 why ?

Comments
Tony says
October 14, 2015 at 11:07 am
Alexa means nothing for backlinks. Try checking in other tools like AHrefs, majesticSEO, etc.

Also, consider backlinks take time (sometimes days, sometimes weeks) to get indexed in those tools.

Comments
Eric Adie Wibowo says
October 14, 2015 at 1:42 am
Hey, I'm looking like this Book for a week. Thank for share this. Its help me to build a video website.

Wow, love this Book so much. So, I learn something new here; Quality Comment! I will try this one. Thank you ☐

Comments
Tony says

June 22, 2015 at 10:48 am
Glad you liked it 󠀠

Comments
Vijaygopal says
June 20, 2015 at 11:09 am
Wow !! It helped me alot. Thanks for sharing. But I have some doubt. How to know that backlinks are DoFollow or NoFollow.

Comments
Tony says
June 20, 2015 at 2:00 pm
You can download a plugin for your browser to check that automatically. Or there are online checkers too.

Comments
Jesse Hackshaw says
June 6, 2015 at 6:15 am
I haven't done a back linking campaign as yet, however, after reading your Book I think I'll give your method and go and sees the results. Thanks for the tips.

Comments
meenaxi says
June 2, 2015 at 1:25 am
I is great Book for seo of my blog .And it is very helpful for my blog or website .

Comments
alamin says
June 1, 2015 at 10:04 am

this post is really good.i learn a good trick from this artical poat.thank u boss to share this....

Comments
vins says
May 30, 2015 at 12:46 am
i really like d case study which always motivate me nd keep me active......
thanx for sharing nic info...

Comments
Muhammad jamshaid says
May 25, 2015 at 12:48 am
Really very good ideas. i like this Book this is very useful for us. i am very happy to read this Book. you have solved my problem. i want to see my post on the

1st page of Google. thanks a lot for good information.

Comments
hari says
May 21, 2015 at 12:45 am
its very very very useful Book…. it helps me to understand the actual concept of backlinks……

Comments
Anas Ali says
May 13, 2015 at 3:35 pm
Innovative Idea! Love this post, I was really looking for something like this. Thanks Alot for Sharing.

Comments

Edwin Rodriguez says
May 10, 2015 at 3:35 pm
This is just what i needed to read. Thanks so much!

Comments
Emma Onwuka says
May 8, 2015 at 4:29 am
Thanks so much! This is wonderful SEO Book and a powerful link building strategy

Comments
Avitus says
May 7, 2015 at 9:17 am
This was a very informative post. I have already started the backlinking now

Comments
Cloud SEO Sydney says
May 7, 2015 at 2:57 am
Google loves only fresh & unique content and high quality back links. So focus on your website content to lead your business on first page of Google.

Comments
Raju says
May 5, 2015 at 10:45 pm
Finally the right Book for me to start link building. Can you please tell me backlinks from website reviews (though they are paid reviews) worthy and safe ?

Comments
Tony says

May 6, 2015 at 12:27 am
Basically anything that's related to your niche and it's not SPAM should be safe.

Comments
Ukeme Jonah says
April 30, 2015 at 2:51 am
I totally agree with the Author! Google doesnt respect crappy backlinks anymore. Quality over quantity. I did more bookmarking than backlinking and my website studyabroad365.com is faring very good!

Comments
jack says
April 27, 2015 at 2:33 am

Hey thanks for sharing this wealth of information regarding backlinks. Now I realized that always content is the king.

Comments
paytm says
April 26, 2015 at 9:47 am
hi,
Is it beneficial to buy backlinks??
And Please give some information about building them. I will be very thankful to you. ☐

Comments
Tony says
April 27, 2015 at 1:09 am

I gave a lot of info on how to build them in my email subscription. Go ahead and check it out 􏰀

Buying them… well… Google hates that.

Comments
Arsalan says
April 23, 2015 at 2:54 pm
I was tired of searching.. how much… how much.. how much.. But finally got what i was wondering. Such a great research. Appreciate that. Thanks

Comments
Syed shahbaz says
April 22, 2015 at 3:58 am

Yeah off course Quality Content is King! Thank you author for this informative post .

Comments
Dave says
April 7, 2015 at 11:30 pm
Thanks for sharing Servando i liked the video part from youtube.

Comments
bamidele says
April 7, 2015 at 1:15 am
Nice tip bro, it help …m..m..m…m.

Comments
diakui.com says
April 6, 2015 at 11:42 am

That is an awesome post. You have really laid out your points in great detail. Very easy to follow with good content. We should take get hold of this post, read and implement the points outlined.

Comments
Kelly says
April 3, 2015 at 11:12 am
good research Servando, definetly quality is more important, hope to see your experiments in 2015.

Comments
Tony says
April 3, 2015 at 1:55 pm
I definitely need to do this experiment again!

Comments
Divankar says
March 30, 2015 at 11:48 pm
nice Book . glad to see this Book ,

Comments
Wiro says
March 30, 2015 at 8:08 pm
Great Book, great website. Thanks for posting this, Servando!
I learn a lot from your Books.

Comments
William Smith says
March 30, 2015 at 3:13 am
hi servando,

Ranking better in google without backlinks is so interesting, to see such kind of innovation in ranking.
Let me try and see.
thanks for wonderful posting, Keep going.

Comments
Lisa says
March 28, 2015 at 6:00 pm
Amazing thing about this post is, the whole thing ends up within the last paragraph that do not start buying thousands of backlinks rather keep your attention in creating high quality backlinks. The question in the heading is always asked by a lot of people and this post will help a lot of us.

Comments
Tony says
March 30, 2015 at 11:53 pm
It is indeed ☐

Comments
verval nrg says
March 27, 2015 at 11:05 am
thanks very much for explain about backlink

Comments
Tonmoy says
March 21, 2015 at 12:59 am
Servando, Thanks for the complete guideline of Backlinks. But i have one question, if i want to do Post Backlinks, i

mean, don't the home page, i want to do Post backlinks to post ranking. Will it works ? Please let me know asap

Comments
Tony says
March 21, 2015 at 12:13 pm
Yes. It works although those have less authority.

Comments
Tonmoy says
March 21, 2015 at 3:24 pm
Thanks for your quick response. So, should i start with this method ? As well, i am newcomer ! What do you think ? What will better for me ? Do to Home

page Backlinks or Do each post backlinks ?

Comments
Tony says
March 22, 2015 at 12:02 pm
Do both to make them look organic as much as possible.
If you can get a home backlink, it means your site is super relevant to the site that's backlinking you.

If you get post backlinks is mostly via news or guest posts. You need both.

Comments
Ged Ward says
March 18, 2015 at 8:10 pm

Hey Servando what other products are available for tracking your website that are pretty useful

Thanks

Ged

Comments
farhan says
March 18, 2015 at 3:27 am
very nice Book i like it thanks for sharing

Comments
Michael says
March 10, 2015 at 3:45 pm

Good experiment but we could expect what results you achieve. Anyway I learn a lot. Thanks.

Comments
chris says
March 10, 2015 at 5:19 am
hi silva
Nice post, I really enjoyed the experiment and learnt a few things.
Do you intend to update the post? ▯

Comments
Tony says
March 10, 2015 at 6:08 am
Hello Chris.
I need to do a new one for 2015. Glad you liked it.

Comments
chris says
March 10, 2015 at 6:13 am
hi silva,
thanks ☐ keep us posted ☐

Comments
alexander says
February 7, 2015 at 11:04 am
hey i have question for you guys so i did add my website url to about 8 website witch is related to my company site but when i check trough seoquake its shows me that i still have 0 backlinks , my question is why i have 0 ? and does it need more time to upgrade the info about my backlinks ?

Comments
Tony says
February 8, 2015 at 9:18 pm
Hello Alex.
Yes, backlinks take some weeks to appear on most backlinks checkers. Try again after a few days.

Comments
Classicchidy says
February 3, 2015 at 7:05 pm
Interesting Book, i have been asking this question to my self several times and finaly i found the answer here. Thumbs up

Comments

sandeep says
January 29, 2015 at 2:25 am
Thanks for putting together this wonderful piece of information.

But what I am missing here, is an RSS feed. After reading the Book, I searched for the same on whole page but did not find it.

I think you should have one and let me know once you have it.

PS: Just started to read Stream SEO Guide and I am loving it. ☐

Comments
Tony says

January 29, 2015 at 1:01 pm
Do you use Feedly or pocket? I think you can add Stream SEO with those ☐

Comments
sandeep says
January 29, 2015 at 1:43 pm
Thanks for the message.

Yes, I do use pocket but still I shall suggest you to have feeds. Believe me, you shall note a significant improvement in your readdership.

Comments
Tony says
January 29, 2015 at 2:20 pm

Thanks for the suggestions man. I had feedburner but that's it and it's gone for a while. Any recommendation?

Comments
beny says
January 28, 2015 at 10:05 pm
im always wondering about this. how much how much ?

Comments
yogesh says
January 28, 2015 at 11:24 am
Hi Servando ,
Awesome study in Backlinks. thanks for sharing this with us. it will really help us

Comments

Ged Ward says
January 21, 2015 at 12:52 am
Great Book Servando. Really enjoyed reading and has given me some ideas of my own to incorporate

Comments
denisrock says
January 19, 2015 at 10:39 am
Hi Just one month back i enter into the SEO world. Presently I confused of Backlinks? Can you explane what are backlinks and how to use for websites these backlink and how to improve site to get more traffic...? These are all my Questions? Please Comments ...

Comments

Tony says
January 19, 2015 at 11:29 am
Hello Denis.
Backlinks are a basic core of doing SEO. I won't explain it here, but I'm sure a quick google search will help you understand how they work and what are they! ☐

Comments
James says
January 9, 2015 at 6:10 pm
I have a question, it might be strange though because i have a website ready and i'm new to seo. The question is that what really are baclinks ?
Is that i make a ad and post it on classifieds with a url link of my website

or post a comment on blogs with a url link of my website ?
or is there a special way to put my url in it to make a back link ?

Comments
Tony says
January 10, 2015 at 2:35 pm
Hello James. Looks like you're starting in this SEO world. I don't have an Book about the basics of backlinks, but I'm sure you'll understand it better after a quick search on Google. Let me know if you've got more questions after that.

Comments
Shajee Fareedi says
January 8, 2015 at 7:10 pm

Its Brilliant, really extremely appreciated. you have cleared my concept and now i have answers for the questions i had in my mind !

Thank you so much !!

Comments
Tony says
January 8, 2015 at 8:29 pm
Thanks Shajee! And thanks for the feedback too. It's been a few months and I didn't notice it!

Comments
Terrance says
January 5, 2015 at 7:47 am

Great Book dealing with back links, I find myself having trouble with creating good links to my site keep up the good work.

Comments
kapil says
January 3, 2015 at 1:14 am
Hey..Your Book is awesome... I red ur Book and i found that you have suggested a blog on seomoz.org but i recently discovered that the company has shifted its domain to moz.com
So if can update the links and mention my blog url that will be grateful... nonetheless you wrote awesome Book which helped me a lot..Keep writing...

Comments

Bilal says
January 1, 2015 at 12:52 pm
Thanks man hope you rank my website

Comments
Richard Gilbert says
December 5, 2014 at 7:58 am
Looks like the amount of back links you have to your website, is becoming less and less important. I did here Google removed back links as a ranking factor as a trail to see what happens but the results were just rubbish. So they are not there yet but i do feel one day they wont matter at all.

Comments
Tony says

December 7, 2014 at 1:23 pm
Yeah. The tests they did without backlink were really bad, but maybe some day...

Comments
paolo says
December 3, 2014 at 5:58 pm
All you say it's true man. Nevertheless I have experienced that I have hit # 1 on google.com for the key word "hair loss and obesity" which has 2,720,000 results, without backlinks: just with good original content, good seo on page, and a web site built with a silo structure.
It's really weird.
I have a relevant (not mine) you tube video enbedded on the page, and a "how to" section within the Book.

You can look it up for yourself searching in google.com "hair loss and obesity" or similar LTKW.

Clearly, if you let me, I will leave my link to this page, so you can recognize it (and some good old backlink building – if you let me :):

I would like to have your opinion on this

Comments
exhug says
December 1, 2014 at 11:20 am
Thanks man hope you rank my website

Comments
Website Design Nottingham says
November 26, 2014 at 7:05 am

Great Book, this information you have shared in priceless. Do you have any more Books like this one, maybe more on were and how to find only the best follow links.

Comments
Sumit says
November 22, 2014 at 9:14 am
I found this very important. Daily i make 20-40 links for my website and all was do-follow and quality backlink.

Comments
iNaVB says
November 17, 2014 at 5:52 pm

Thanks for the information Servando..
This Book very help me to explore my site 🙂

Keep update.

Comments
German says
November 17, 2014 at 3:01 am
Hi Servando,
very interesting experiments.
I also wonder what tool do you use to track your position, the one you attached the screenshot from?
thanks

Comments
Tony says

November 17, 2014 at 11:03 am
Hello. I'm using ProRank Tracker. They have a free plan for a few amount of keywords. Check it out!

Comments
Zebi says
November 18, 2014 at 2:14 am
Thanks a lot for this effort, disclosing experiences let others learn quickly and give them valuable insights. keep up the good work

Comments
n rawie cumaar says
October 30, 2014 at 9:53 am

Excellent Post! superb information will help me to get quality backlinks rather than the quantity

Comments
Jessica Banks says
October 30, 2014 at 3:35 am
Thanks for such a informative Book. I am using similar strategy for my website on Master data Management

Comments
Marilyn says
October 29, 2014 at 4:33 pm
It's always great to hear about SEO experiences. Great Post.

Comments

Muhammad Abdullah says
October 21, 2014 at 12:26 am
Though i have already created more then 10k links for my new website, but your argument is logical. from now i will prefer quality too, Thanks for sharing!

Comments
Dedy says
October 16, 2014 at 11:56 pm
Thanks for sharing these tips on backlinks. I'm just now getting started in this area, and you provided alot of vital info.

Thanks,

Comments

Mono says
October 16, 2014 at 12:38 am
This is a good information for me as a beginner, still learn a lot of information like this. Thank you for sharing this Servando..

Comments
Dani apri says
October 13, 2014 at 1:12 am
thank for sharing, it is very interesting for me as new commer on SEO

Comments
surya says
September 23, 2014 at 2:06 am
how to reduce are website loading time if images are very big or big resolution

images we are using. because it will effect our bounce rate.my website having 70% to 80% bounce rate how i will control it.Gzip is important for SEO plz tell me the process for gzip how i will do gzip of my website.

Comments
Lekan says
September 7, 2014 at 12:19 pm
This is one of the best I have seen on Backlink building strategy. Great minds are not made by age, but by wisdom. You've got the wisdom! Thanks for this. I hope I have known this for a couple of months back.

Comments

Vaibhav Gupta says
August 30, 2014 at 9:17 pm
How can i create google and yahoo as a back link to my website?
I have seen many websites in alexa directory with google, fb, yahoo as their links.. How can I implement the same? any ideas?

Comments
Tony says
September 1, 2014 at 11:16 am
Maybe you're confused. You won't get backlinks from google or yahoo pages unless you're approved on google /yahoo news. From yahoo, you can also get backlinks from yahoo answers. FB

links are no follow so they pass now Page Rank juice.

Comments
Enumah Chinedu says
August 27, 2014 at 4:19 pm
You are indeed a technology master…..nice one bro

Comments
Parth Kheni says
August 19, 2014 at 4:10 am
Hey,

Thanks for this Book. Now I understand what I need to do for my Book to view Google's first search result page.

Thanks again.

Comments
lata says
August 7, 2014 at 12:04 am
Hello
I want to increase my website on the top of the Google first rank , i make more efforts but it does't work.
Can you suggest me a best technique for this.
Thanks for sharing.
For more details , please visit on this link.
jonsonsglobal.com/
Thanks.

Comments

Tony says
August 7, 2014 at 11:40 pm
Hello. Have a look at the guide I have for my email subscribers. There's good info there to start ☐

Comments
Al-Amin Kabir says
August 4, 2014 at 1:17 am
Hi, Do you use your exact post link instead of using the homepage link while commenting? Is it good practice? What do you think?

Thanks for the case study, it's amazing!

Comments
Ajay kaushik says

July 21, 2014 at 1:08 am
Hello Servando ,
Your Book Is realy Intersted. So Please tell me best way to get do follow backlinks.

Comments
Tony says
July 21, 2014 at 9:37 am
Hello Ajay. Your best bet for high value do follow backlinks at this point is to build PR backlinks. I have a complete guide on how to do it for my subscribers

Comments
Parth says
July 14, 2014 at 11:57 pm

Nice post . Sir this means we need just few backlinks but quality ones?? How to get that??

Comments
Tony says
July 15, 2014 at 11:26 am
Yes.
Yu can do guest blogging or build your own links as explained in the guide form y subscribers ☐

Comments
Lopeholt says
July 14, 2014 at 7:02 am
One simple question! How did you create those backlinks via blog

comments? Did you use your own name or the Keywords in place of the names? Great experiment. I'll give it a try.

Comments
Tony says
July 14, 2014 at 11:52 am
No, I don't like to look that SPAMMY. I used blogs that have commentluv or similar systems enabled back at the time (those allow you to put a link to your blog at the end of the comment).

Comments
Justin says
July 13, 2014 at 11:49 am
I noticed your blog today only. The topics are very useful especially this post

help me a lot. Many of my doubts are cleared from this post. Thank you for sharing this great post.

Comments
Tony says
July 14, 2014 at 11:46 am
Thanks Justin. Glad you liked it.

Comments
Marsha Flavour says
July 10, 2014 at 3:58 pm
This is a really great experiment i think i am going to try it because sometimes i build hundreds of back links with little success so maybe this will work for me.

Comments

boxer says
July 10, 2014 at 5:02 am
great artical bro. backlink is powerbank of a blog.if don't have a blog any backlink it's not getting high rank on google.i will try to making new backlink for my new blog.thankyou very much for sharing valueable content

Comments
Dermot Gilley says
June 27, 2014 at 9:34 am
This is a fascinating, exhaustive and exhausting story. Thank you for sharing and for taking the trouble in writing it up in so much detail (most don't bother publishing the winded path to success, which is one of the reasons many

rookies come to think it must be easy). "All these backlinks appearing on my rank tracker are backlinks created by commenting 1 month ago or so on January/December." – That one I don't quite understand, though: Does backlinking via comments really benefit SEO? Mostly it leaves nofollow links in its wake. We're told that nofollow links don't increase PageRank. Most blogs only create nofollow links for comments. These would be useless as an SEO tool since the nofollow attribute avoids leaking PageRank. Wherever I look this mismatch between the proposed benefits of backlinking via comments and the downgrading of these backlinks by virtue of the nofollow attribute is never discussed. If there was an easy solution to the puzzle, everyone would talk about it. So I suspect that many authors are groping in the dark. In your case it seems to work astonishing well and I am wondering why.

Comments
Tony says
June 28, 2014 at 11:04 am
Both backlinks help. Do follow or no follow, but some of them will increase your page rank and some others only help to have a diverse anchor text link report.

Comments
dunia remaja says

June 5, 2014 at 11:10 pm
Thanks for this.
I liked it. I want to know some thing that Can i make comment on different Books of same blog for generating many backlinks to increase ranking. Will this trick will incease my ranking in google?

Comments
Tony says
June 6, 2014 at 2:30 pm
It will help you get more backlinks and different anchor text, but it's much better to get different backlinks from different blogs instead of the same blog within different pages.

Comments

Chad says
June 1, 2014 at 1:21 pm
I am new to the SEO world. I have sat around and thought about it and talked myself out of getting into blogs for years now. This year i finally decided to get into it. I really appreciate Books like yours and information that other people are providing for free on a website without spamming us.

Comments
Tony says
June 6, 2014 at 2:04 pm
No problem, Chad

Comments
Vjay says

June 1, 2014 at 2:26 am
Awesome Book, you have applied wonderful tactics to rank your blog on Google.

Please keep up updated with such type of SEO Tips, so that it can help us also

Comments
Tony says
June 6, 2014 at 2:04 pm
Glad you liked it. I have one new Book coming, so make sure to subscribe ☐

Comments
Brahim says
May 29, 2014 at 2:54 am

This is very interesting, Servando; exactly the ufo i was looking for , thanks for sharing

Comments
ana haberler says
May 26, 2014 at 2:07 am
BacLink much needed. I thank you for the information you provide.

Comments
sam says
May 24, 2014 at 5:16 am
very nice content keep it up.

Comments
Mohammed Basheer says
May 23, 2014 at 5:00 pm

Hi...dude...am totally impressed with your Book

Comments
Mohammed Basheer says
May 23, 2014 at 4:59 pm
blog looks very interesting..great content

Comments
Ed PH says
May 23, 2014 at 4:42 am
Thanks alot Servando, for a moment i thought backlinking is dead. Have to start working on quality backlinks now. Good job there.

Comments

Mohsin Khan says
May 21, 2014 at 8:27 am
hey Servando, can i get backlinks from niches other than my niche ?

Comments
Tony says
May 23, 2014 at 5:17 pm
If the backlinks are going to be relevant, yes. Otherwise I wouldn't recommend it.

Comments
Mohan Joshi says
May 7, 2014 at 10:41 am
Best this you have dicussed is that many people are not sure that backlink are not working but its not true. this is being

observed in your art that it works many more for google ranking.

Comments
Richard Orchad says
May 7, 2014 at 3:55 am
My forexsanity website is 6months old and I have 96 backlinks already. Many of these backlinks are coming from forex forum. Are forum links strong enough to give me page rank? How many more links do I need to get to page rank 9/10 or 2/10. Please share

Comments
Tony says
May 7, 2014 at 7:34 pm

Forum links are good, but probably won't pass enough link juice to rank you. Specially if they all come from the same forum.
Since forex is a very difficult keyword, I'd say it will take a good SEO campaign for it.

Comments
vikraam says
May 1, 2014 at 7:30 am
hey in one Book can i have backlinks from 3-4 different index of a same website?

Comments
Tony says
May 2, 2014 at 11:10 am

3-4 backlinks from different indexes? Can you give me an example?

Usually, 1 backlink from each domain should be enough. It's better to have backlinks from more domains instead of multiple backlinks from the same domain.

Comments
Zethora says
April 28, 2014 at 8:00 am
Thank you for sharing this useful information. Quality is certainly more important now and makes sense.

Comments
khalid says

April 23, 2014 at 6:59 am
thank you for sharing its really helpful. keep us updated.

Comments
Kidd Activity says
April 18, 2014 at 11:10 pm
Would it be safer to maybe make 5 links to different Books then to make them all go to your domain each day? I don't want to get the boot by Google!

Comments
Tony says
April 19, 2014 at 12:05 am
That would work if the Books have high authority.

Comments
forumfusion says
April 5, 2014 at 10:33 pm
How many back links do i need?
before reading this Book this question is bomb in my head but now my headache is gone.Thank you for really awesome information. ☐

Comments
Tony says
April 6, 2014 at 12:20 am
It's great to know you liked it ☐

Comments
Jake Makere says
April 1, 2014 at 3:27 am

Thanks for the Book. This helped me a lot, and I especially agree on the dofollow backlinks. Hard to come by though.

Comments
grace says
April 1, 2014 at 5:01 pm
That you for this info on backlinks, i thought we needed 100's, you saved me a lot of headaches !

Comments
Tony says
April 1, 2014 at 11:31 pm
Hello grace.
It depends on que competition and the quality backlinks.

If you're building low authority and totally unrelated backlinks, you might need hundreds or even thousands. But a few quality backlinks can do the trick as long as they're relevant.

Comments
Drafting says
March 30, 2014 at 8:18 pm
Whoops! Didn't see that there.. ☐ Thanks for getting back to me.

KR.

Comments
Tony says
March 30, 2014 at 10:40 pm

No problem 🙂
Glad I could help 😊

Comments
Prerna says
March 30, 2014 at 1:50 pm
I agree acklinks should be from quality pages but does it matter that a backlink shoud be a dofollow one and nofollow links are of no help.

Comments
Drafting says
March 30, 2014 at 6:48 am
Interesting ideas.. So when's the next update Servando?

Comments

Tony says
March 30, 2014 at 8:15 pm
My 6 months results have been updated and you cna find the link to the Book at the beginning of this one ☐

Regards!

Comments
BHM Labs says
March 24, 2014 at 10:52 am
Hello Mr. Silva , Your Book was very nice. I am totally new to this SEO world , So I am going to ask you many questions only if you don't mind.
Here goes my questions.

1. How to get quality of Backlinks.

2. How to increase the ranking for a particular keyword.

These are the questions which are in my mind from a long time. Hope you will Comments me and help me in getting a good knowledge of SEO.

Comments
Tony says
March 24, 2014 at 1:33 pm
Hello there.
1. Quality backlinks means they're relevant baacklinks from old/stablished websites that have tons of Books, traffic and Page Rank/Domain Authority. So, there more you get of those, the better.

2. You need to do a lot of things including backlinks, Social media signals, getting traffic and in page SEO.

It's kind of difficult to explain it here, but have a look at my SEO/Marketing guide to start right away!

Comments
Cathy Dollartrick.com says
March 26, 2014 at 1:41 pm
How many backlinks are safer to be made each day/week ? If I create 8-10 backlinks per day, would it be a safe job so that Google doesn't consider it as spamming ? Thanks for the nice Book.

Comments

Tony says
March 26, 2014 at 2:49 pm
8-10 is OK.
Thousands per day is what Google considers spammy, normally.

Comments
Jospeh says
March 18, 2014 at 1:08 am
Hey Servando,
I was started health blog one year ago. But I have not received enough traffic for my blog. How to build back links for my health blog? Help me

Comments
Tony says
March 18, 2014 at 1:13 am

Hello Joseph.

Have you tried creating Books in other places, guest posting, Book submissions or even comments on other blogs?

You should subscribe to my newsletter because it will help you with backlinking strategies ☐

Comments
Pawan says
March 15, 2014 at 7:06 am
Great Book. Can you help me as to where can i make quality back-links? I would really appreciate the help.

Comments
Tony says
March 15, 2014 at 5:51 pm

Hello Pawan.
There's a free guide for my subscribers on how to build high PR backlinks, if you're interested on it ☐

Comments
Nitish Dubey says
March 7, 2014 at 4:34 am
Hi,
I must say your case study is quite an elaborate example for helping out new bloggers like me. When I started I had a very narrow view of backlinks. Looking at your weekly log, it really threw some insights on the relationship between the backlinks and the search engines. And finally the comments were helpful as

well. Thanks for clearing things up about the backlinks.

Comments
Tony says
March 8, 2014 at 11:23 pm
Hello Nitish.
I've come to the idea that case studies are basically the most useful Books for starters and experts.
I'm glad it helped you.

Comments
domique lopez says
February 25, 2014 at 12:36 pm
Interesting Book and analysis. Im looking forward to see how this experiment goes.

Subscribing

Comments
Alex says
February 20, 2014 at 8:53 pm
Hi, thanks for the informative post, I was wondering if you still use market samerai, and is it still an effective and accurate tool for niche and market research? Can you recommend me another tool if MS is no longer effective.

I would want to eventually be able to research why sites rank on page one, why is that specific market so hot (i want to validate the hot trends), what tools besides market samerai will help me do the job?

Comments
Tony says
February 24, 2014 at 1:26 am
Hello Alex.
Yes, I still use Market Samurai and it's very effective even if it's a little slow.
I haven't tried lots of good tools, but I do know that Longtail pro is good.

Comments
sutopo says
February 20, 2014 at 7:56 pm
thanks your share information,amazing to how many backlink do need..
regards

sutopo

Comments
Printer dtg says
February 10, 2014 at 6:07 pm
Hi Tony, many thanks for your great Book. Do you think paid backlinks will hurt our website? So many company offer us to buy backlinks and i think it is an easy way to get backlinks

Comments
Tony says
February 10, 2014 at 10:17 pm
Paid backlinks are seen as manipulation for Google. Honestly, it all depends on if you leave a footprint or not.

Cheap, bad quality backlinks are probably going to hurt your rank in the mid-long run. Quality backlinks are fine, but you need to check your anchor text ratio and make them natural.

Comments
John says
June 11, 2014 at 2:23 am
Paid backlinks are a bad idea. google will delist a site if they see this happening. Not a good idea for sure.

Comments
Tony says
June 11, 2014 at 11:19 am
Actually. All backlinks are a "bad idea" for Google at this moment.

Or at least that's what Google says.

Comments
John says
June 11, 2014 at 12:38 pm
Really? See I thought backlinks were still important for SEO? There is conflicting data on this around the web though, you know?

Comments
Tony says
June 11, 2014 at 12:41 pm
They are. Backlinks are still very important.
That's why Google doesn't like how people get backlinks anymore. No matter if they are paid, guest posts or

natural links. You can get penalized anyway.

Comments
John says
June 11, 2014 at 2:54 pm
Hi, Servando,

That makes sense. From what I am reading, Google is really paying a lot of attention to user experience and social mentions (i.e. someone mentioning you on Google+.

I have found that Google+ is a HUGE way to help with getting traffic and SEO. The key is being interactive on the platform and doing so consistently.

This is a great topic – so glad you brought all of this up!

achin verma says
February 7, 2014 at 3:51 am
hi thank you for such an informative post. i love to read posts which are based on experience. i want to ask you which tool you used to check the position of your post on different search engines. I'll be wating for your answer

Comments
Tony says
February 8, 2014 at 11:49 pm
Thank you.

I was using Market Samurai, but now I use Pro Rank Tracker 🙂
Hope that helps.

Comments
zeshan says
February 4, 2014 at 4:05 am
Wow that's some really serious backlinking research ! Loving it

Comments
Emily Jenifer says
February 1, 2014 at 4:21 am
Hi Servando,
don't u think creating direct backlinks to your main URL without any tiers downward will hurt in long run?

Comments
Tony says
February 5, 2014 at 2:38 am
Hello Emily.
For low quality and high volume backlinks, yes, they could hurt. But high quality or PR backlinks shouldn't be a problem. They are normally the minority.

Comments
kurs says
January 31, 2014 at 9:08 am
i searched google for "how many backlinks needed" and came here. wow, the best Book about backlink building which i ve read until now. this Book is a very talented guide what is the

backlinks' role on serp, how to build backlinks to get higher rankins and more trafic etc. after i read it, i realized that you are a seo guru. best regards...

Comments
Al-Amin Kabir says
January 26, 2014 at 12:09 pm
Great one Tony!

Waiting to see your further post on backlink backlink experimenting ☐

Comments
Tony says
January 26, 2014 at 4:57 pm
Hello.

The latest update was published a few weeks ago and it's already there. Check at the beginning of the Book!

Comments
Frank Spohr says
January 24, 2014 at 2:51 pm
I enjoyed your Book Servando. I too have found that I don't need that many back links to get into profitable search engine rankings. It really is about having a good mix of quality. I had a competitor who literally had hundreds of thousands of back links and I was able to beat him with a much smaller batch of higher quality links.

Comments

Mark says
January 23, 2014 at 4:21 pm
Great Book, lots of good ideas to try and help generate back links. Do you find commentluv is better than Disqus or LiveFyre?

Comments
Tony says
January 23, 2014 at 7:46 pm
I think it's just plain different. Commentluv is basically to create backlinks, and outside of the internet marketing, SEO or MMO world, nobody would be interested on having it, but I could be wrong.
Disqus is great for most of the niches. I prefer Disqus compared to Livefyre.

Comments
Jony says
January 18, 2014 at 7:50 pm
Hi,
Your experiment is promising. I am trying to achieve the exact same thing right now with my 3 month old website. I think I learned a few things from you today. I just do not want to go the volume route of building 1000s of backlinks.
Thanks for the post

Comments
Tiang Beton says
January 17, 2014 at 12:47 am

Thank you... I'm getting smarter... because I read and hearing from you...

Nice to meet you.....

Comments
Manya Singhal says
January 15, 2014 at 11:29 am
that's really great experiment, i read it and got that quality is more important than quantity !

Comments
Harry says
January 7, 2014 at 3:39 am
I need help in building backlinks please contact me through facebook ,
http://facebook.com/haren.mukeshkum

ar . I really need help if you are free please message me!

Comments
Tony says
January 7, 2014 at 1:11 pm
Hello Harry.
Any help or questions can be solved here. Feel free to ask □

Comments
julian says
January 2, 2014 at 11:17 am
Where do we find the dofollow backlinks?

Comments
Tony says

January 2, 2014 at 2:17 pm
Hello Julian.
There are a lot of ways to get do follow backlinks including Books, hubpages/squidoo/infobarrel, contacting other blog's owners and more. Commentluv also allows that depending on the settings chosen.

Comments
Pain behind knee says
December 30, 2013 at 10:51 am
That's good Book .. I read all of it and will apply it to my blog right now, can't wait to get up in results :
)

Comments

Amar Ilindra says
December 25, 2013 at 7:18 am
This is really interesting case study. I started few niche sites and i hope my site ranks first for my targeted keywords. Your experience will surely help me

Thanks

Comments
Tech Bhuvan says
December 23, 2013 at 11:18 am
Your Book was really interesting and incredibly detailed. I now know how to play with creating backlinks to my blog Books. I was wondering if backlinking at all works in search engine ranking. And now I clearly know it does, but we do

not need thousands of backlinks to get listed in first page of Goolge. Nice post indeed. Keep informing us with such informative ideas. Thanks a lot.
Tech Bhuvan

Comments
Aaron says
December 19, 2013 at 1:17 pm
Hi, I love your Book. I'm going to try it and since I'm battling with two other guys on the SERPS I must do this. They are killing me with their 1000's of back links in a few weeks. But, I will get them.

I'm glad I do real reviews.

Comments

Ron Terrie Soriano says
December 18, 2013 at 2:46 am
Thanks for this Book! I really want to try that experiment to my site.

Comments
Daniel Wu says
December 17, 2013 at 3:12 pm
Hello I was just wondering but how would you make quality backlinks for youtube videos?

Comments
Tony says
December 18, 2013 at 9:44 pm
Hello Daniel.
I'd actually follow the same strategy as if it was any other website's page.

The link, the right anchor text and diversify!

Comments
burs sorgulama says
December 15, 2013 at 3:44 pm
nice info thanx i will work on it

Comments
Matt says
December 13, 2013 at 12:53 pm
Thanks for this post! It's amazing to me how many of my clients still believe in black-hat techniques and directories to get high search rankings. On-page optimization (titles, headers, etc) and quality backlinks are the way to go!

Comments
Jenny says
December 13, 2013 at 11:49 am
Great research, I was wondering why I only had 2 back links in Google and not more.

Comments
Tony says
December 13, 2013 at 12:34 pm
Hello Jenny.
Where did you check that information?

Most sites take some time to index the backlinks. Some of them are updated every month, so please be patient and you'll see results!

Comments
Amy says
December 12, 2013 at 6:28 pm
Great Book. I wonder how it works for my Etsy Shop Artfanaticus

Comments
Tony says
December 1, 2013 at 6:36 am
Great Book! Thank you for sharing it.

Comments
Webentwicklung says
November 29, 2013 at 6:38 am
I like your approach. And might try it to for another project I am working on. Documenting it will also help to show proof to our clients…

also I would like to add that we have a couple of Books that rank very well on google.com and these were well researched Books about how to solve a technical problem. As a tip in general. If you come up with something other people might like, the share those links and repost them... we get new backlinks every week and don't really have anything to do with that. it's mainly just about writing good content.

the only problem in the beginning is that if you wrote an articel you got to promote it and that often is seen as spam. but i guess it depends if you put the comment into related blogs or not.

Comments
Sekhar says
November 27, 2013 at 7:21 am
Great research man, I was really fascinated you did all these research. It gave me great insight into how backlink and ranking work in relation. I was looking for how backlinks are related to help raise page rank and bang on. Your post was what I am looking for. One small question just for my help, how to use anchor text while commenting?

Comments
Tony says
November 27, 2013 at 10:54 pm

I really don't recommend using anchor text for comments. They look really spammy. Instead, you can check blogs like mine that have commentluv enabled and they allow you to drop a link to an Book without looking like a spam bot.

Comments
Ankur Upadhyay says
November 25, 2013 at 12:27 pm
Hii ,

You have done a great research on backlinks. I am trying to rank on the first page for my keywords. I will try to follow your lead. Looking forward to your updates,

Thanks for sharing this awesome Book. ☐

Regards,
Ankur

Comments
Tony says
November 25, 2013 at 1:56 pm
Thank you Ankur. I'm currently not updating this one anymore, but if you watch at the top of the Book there's a second Book for a follow up on what works and what doesn't after a few months. That one is updated ☐

Comments
yohan says
November 21, 2013 at 9:34 pm

Nice introduction, I want to adjust my website according to you and follow your steps.

Comments
Ian says
November 17, 2013 at 12:58 pm
Hi,
So, i want to create some awesome backlinks, but commenting of different Books and leaving my website may not be good enough, i want legit backlinks to my site and share what i have! Please help me guys!
Cheers

Comments
Saqib Rauf says

November 17, 2013 at 9:19 am
Got your point. Can you tell me, how can i find blogs that gave me dofollow links?

Comments
GK says
November 8, 2013 at 3:50 am
Wow that's some really serious backlinking research ! Loving it ! ☐

Comments
Bisnis Online Terbaik says
November 3, 2013 at 11:47 am
Thank you.
I have a nice update for this week. Going up all the way faster than I thought.

Comments

Ryan says
October 30, 2013 at 3:33 am
Don't forget the that the content is the most important thing!In my opinion buying backlinks is a bad idea.

Comments
mahmoud eladawy says
October 27, 2013 at 10:08 am
Hello my friend,
I do you advice, but i have a question :
I have my blog and have a competitor who get his backlinks for un relative sites as :Skype community and Kasperky and Bitdefender in the other side i get my backlinks from relative site and relative content and his rank rise very fast but my blog slowly.

Comments
Tony says
October 28, 2013 at 12:39 pm
Hello Mahmoud.
If your competition is getting high PR backlinks like the ones you said, why don't you copy them? I guess they're building forum backlinks, right?
What's your plan for that? Prepare a plan and do it! If you already know where are they getting their backlinks from, you've already got the weapons. What's stopping you?

Comments
gikoberry Incorporation says
October 25, 2013 at 10:16 pm

This information is really worth. I haven't tried social bookmarking. its a must to try!
Thanks for sharing.
—

http://www.gikoberry.com

Comments
Vaibhav says
October 25, 2013 at 1:18 pm
I found this Book interesting, a week by week experiment. I don't have to create as many backlinks as I thought, just focus on quality 2-3 per week looks so promising.

Comments
clement says

October 17, 2013 at 4:40 pm
Just started two new blogs and the info i got here will be very helpful. Thanks so much for sharing and keep the good work flying.

Comments
Tony says
October 17, 2013 at 4:50 pm
Perfect man. Let me know if you have any questions!

Comments
trendy prom dress says
October 16, 2013 at 7:33 am
i am asking the same question , that's why i find this blog on google , i am doing the backlinks for my own website ,

i've learned a lot from your Book and thanks for walking us step by step

Comments
toufan says
October 12, 2013 at 6:29 pm
wow, your experiment is great,,,,i think 1-3 high quality backlink every week is more better than many backlinks but not quality,,,,thanks for sharing

Comments
Tony says
October 14, 2013 at 4:02 pm
1-3 High PR backlinks sounds great, toufan. And definitely doable.

Comments

tob says
October 12, 2013 at 10:29 am
Thanks just opened a blog and all this backlink has been getting too much for me but RBIs would help me focus on quality

Comments
Watesh says
October 9, 2013 at 10:13 pm
That is some really good research. It has actually helped me with lots of amazing information

Comments
Patrick Shea says
October 6, 2013 at 9:10 pm

Very nice read. I have learned a lot here. I will try this out myself.

I also struggle with Back Links and trying to get on the first page of Google. Its no easy task.. But reading what was said here will help me with that goal. Thanks for the good post.

Comments
Tony says
October 7, 2013 at 5:12 pm
I'm glad it helped you. Now go and build some backlinks!

Comments
NearPage says
October 5, 2013 at 12:37 pm

how can i see which place my website is shown in google??

Comments
Tony says
October 8, 2013 at 11:51 am
You can use any rank tracker like Market Samurai or get one for free if you look in google.

Comments
Megan says
October 1, 2013 at 1:41 pm
So, how long did it take you to build each of those back links? I am new to the promoting but not to blogging. I actually started out on Hubpages.

Comments
Tony says
October 2, 2013 at 2:53 pm
I used hubpages a few years ago too. How long? well, every update I point out here is spaced between 1-2 weeks, so that's how long it took me.

Comments
MyBlogMyEarning says
September 27, 2013 at 9:13 pm
Quality Of Backlinks Are Much important Then Quantity of Backlinks So Everyone Should Concentrates on Quality instead of quantity.Anyways A Very Nice Post Keep It Up

Comments

Toni91 says
September 24, 2013 at 9:44 am
Don't forget the that the content is the most important thing!In my opinion buying backlinks is a bad idea.

Comments
Anirudh Sharma says
September 20, 2013 at 6:56 am
If you are having 4,100 backlinks ….then why is your Page Rank 0…..?

Comments
Tony says
September 21, 2013 at 9:23 pm
Actually, I have around 19k backlinks now (majestic SEO source) and my PR is 1.

it's been one since the beginning of the year when I had those 4k links, but Google hasn't updated their PR lately. I guess I could achieve at least PR 3 now if they updated it. But I don't care anymore as my Citation flow (authority score) is higher and this Book has good rankings anyway.

Google is moving more into a Social/authority signals search engine and haven't cared about updating PR since February 2013.

Comments
Karen says
September 19, 2013 at 5:11 am
Hi Servando

Can you give us any insight to whether or not back-links from a quality forum are of any value or does Google only count one or two links, either from the profile or the signature?

I think forums are more about building a brand rather than back-links (or at least they should be) but I am concerned that many businesses might not join a forum anymore as they are worried about the sheer amount of back-links which would be generated from the use of signatures. It seems that this idea of multiple back-links from any single domain might harm a business site?

What do you think or does Google count only so many back links and disregard the rest?

Comments
Tony says
September 21, 2013 at 9:26 pm
Hello Karen.
Backlinks are almost always valuable, and Forum backlinks (no-follow or follow) are always valuable too.
The thing most people thing that because they have 500 posts in a forum they will get 500 backlinks instantly and get ranked, but it doesn't work like that.

Google measures the number of backlinks coming from the same website

and even the same IP, and if they see a lot of them their value is usually decreased.
So it's better to have 100 backlinks from 50 different websites, than having 100 backlinks distributed in only 10 websites. And the PR also helps a lot.

Comments
Elizabeth says
September 2, 2013 at 7:10 pm
This is a great case study. I constantly preach quality of content to my clients. Thanks for sharing your SEO insight – really great!

Comments
Tony says

September 8, 2013 at 1:51 pm
No problem, Elizabeth. I'm glad you liked it.

Comments
Cricke says
August 29, 2013 at 9:42 am
Nice artical You changed my mind I was that Anchor text should be with your keyword at least 80% not you have proved me wrong thanks for correction

Comments
moniways says
August 29, 2013 at 11:06 am
This Book is great. I bought 10000 backlinks from fiverr and up till today, i

have not seen any improvement on search result. Thanks for this.

Comments
Tony says
August 29, 2013 at 11:50 am
Yeah. That's usually a bad idea. Either doesn't work or gets you penalized.
No problem ☐

Comments
Tony says
August 29, 2013 at 11:52 am
Anchor text on 80% of your keywords will definitely get you penalized nowadays.

It was one of the main reason for google to make updates in 2012. Keep it below 20-30% if possible.

Comments
Gautam says
August 28, 2013 at 6:39 am
New to your blog & really like what you have share. You have clear my many doubts about Back links backlinks. Thanks for sharing

Comments
Tony says
August 29, 2013 at 11:59 am
No problem man. Hopefully you'll subscribe to get more updates!

Comments
ige ebima says
August 21, 2013 at 12:21 am
Thanks for this lovely and useful experiment,i know know what do do to rank well

Comments
Pankaj Jain says
August 16, 2013 at 6:44 pm
Hello silva,
Thanks for this Book, you have cleared many doubts about backlinks. Now I know what I need to do for my Book to view on first page of Google.
Thanks again.

Comments

vanaja says
August 16, 2013 at 1:32 pm
this Book is certainly a helpful one

Comments
Vanessa Chrales says
August 16, 2013 at 4:53 am
Hi Servando. I really appreciate your experiments. Now major search engine like Google gives priority to quality and same niche links. Very nice case study. Keep sharing like this with us. ☐

Comments
Tony says
August 18, 2013 at 3:43 pm

Thanks Vanessa. Sure I will keep writing more posts like this. Case studies are the best!

Comments
bababato says
August 13, 2013 at 12:18 am
Good day Servando agreed with your post the links which show more relevancy, near to theme and appear natural to search engine have large life then dumb ones ….Keep updating

Comments
MoneyMazics says
August 12, 2013 at 10:27 am
Seems amazing but building back links is quite uphill task..

Comments

Hormigon Impreso Sevilla says
August 9, 2013 at 7:30 pm
Hey Sevando, after your 10-30 quality backlinks, try to get 3-4 edu and 3-4 gov comments, and 50-100 facebook shares, likes, +1, pins and twitter share's to your Book, You must make him "viral" in the eys of the search engines. You'll have your No1 place after 1-2 weeks!

PS: my EMD's are going perfect. I have over 100 EMD's, and I was not penalized by mr. G!

Comments

Tony says
August 12, 2013 at 12:19 pm
Hello, thanks for your recommendations. Actually, I already had around 50 interactions (likes, tweets) and right now my Book is ranking number 1 again. It happened a week after I published this post or less.
About EMDs... yeah. They're still rocking.

Comments
Money says
August 13, 2013 at 9:26 am
Am not getting back links and I feel that most of the people are buying them..

Comments
Smart Seo says

August 2, 2013 at 6:55 pm

This is a nice Book. Nice and Informative work. But I expect more broad details about how many backlinks can get a high competitive keywords(12000 local Monthly searches/month) to get first page in Google. I expect an informative Comments…..

Comments

Tony says

August 3, 2013 at 2:14 am

I'll work on something like that on one of my future Books. However, even with long tail keywords receiving just a thousand search queries per month, you can make some good money.

Comments
Raitis says
August 2, 2013 at 6:06 pm
Hi! This is realy helpfull expreiment. I have a site with financial loans collation, what have an zero pagerank. I already made some 20 quality backlinks like you described in your expreiment (one till three backlinks in week with different rext), but now is around month spended and it still zero pagerank. Any sugestions, ok I understand – just patient.. Thanks

Comments
Tony says
August 2, 2013 at 6:12 pm
Hello Raitis. Yeah no problem.

Google only updates the Page Rank every some months. It's been 6 months already without an update. You can also use Market Samurai to check Citation Flow, which actually updates in real time.

Comments
daydaily.com says
July 24, 2013 at 8:48 pm
thank you very much for the information

Comments
J. Michael says
July 24, 2013 at 1:56 pm
That interesting. I'm in the process of getting as many backlinks as I can. First Iread the numbers count and then I read

that quality counts more. In the end maybe I'll have it figured out. I think for now I'll do both.

Comments
Tony says
July 29, 2013 at 8:58 pm
Hello Michael. Both count, but quality is superior than quantity at this moment, and it looks far less SPAMMY too.

Comments
Roach says
July 19, 2013 at 4:27 am
Oh wow, very useful and taught me a lot about how to do this and such! I will be following this guide and bookmarking for

more and more SEO tips. I have found SEO heaven!

Comments
Tony says
July 20, 2013 at 1:47 pm
Thanks Roach! I'll publish my second part on this backlinking experiment soon (this week). I've found very interesting things.

Comments
Clark says
July 3, 2013 at 3:01 am
Hi servando, great information. Information that you have provide is quite interesting and informative.

Additionally, give the inspiration for new blogger. i would surely be try it.

Comments
Tony says
July 9, 2013 at 7:38 pm
Thanks Clark. I'm glad it helped you!

Comments
Arindam Dutta says
June 25, 2013 at 6:21 am
This is great and will be very helpful. Basically few backlink from great sites are better than hundreds backlink from lower sites. You can always gain high quality backlinks with social bookmarking or if you submit some

Books to high page rank submission sites.

Comments
Vijesh says
May 17, 2013 at 1:03 pm
Nice and clear explanation of how quality back links help in ranking top in Google search results.
Thanks for the share Servando, need to learn such SEO techniques to make my blog rank top in search engines.

Comments
Tony says
May 17, 2013 at 2:15 pm
Hello Vijesh. Thanks for your comment.

I'll write an update about this on a separate post next week. Many things have changed the last month but I had a really hectic month doing some business and preparing some stuff. Still, I think you'll like it.

Comments
Suresh Mothasara says
April 21, 2013 at 11:24 pm
nice post bro. backlink is a power of post and i was taste the backlink a blog have higher pr to make lot of backlinks so i have just started to create backlink to guest posting on my niche sites oyur post is very nice so thankyou very much

Comments

Mike says
April 14, 2013 at 4:36 pm
This is a fantastic experiment – the level of detail and analysis – you really break it down. I've shared in on blogbods so newbie bloggers take note.

I'm amazed at how few backlinks you can get away with and still rank. Of course this is all down to quality content on your part. We're not going to get the same results unless we focus on the quality (and quantity) of our posts.

Fantastic stuff – thanks!

Comments
Tony says

April 15, 2013 at 12:53 am
Hello Mike.
Thanks for sharing it. I actually have to update it because there have been some interesting changes the last 2 weeks I haven't documented yet.
I just need to get some time and I'll do it.

Comments
jeremy mcdonald says
April 8, 2013 at 4:18 am
if you have something new to offer google will praise you immediately without any delay ...you need not an exact number of back links to get on top you just have to work on the stuff which you want to index better on search

engine a little bit creative and unique thanks for a great share

Comments
Tony says
April 8, 2013 at 10:30 am
Hello Jeremy.
I'm with you, and using Google trends is a good way to achieve that.
However, not always can we put a totally new Book in this crowded world. So you still need to make better things tan the competition.

Comments
Tulisa says
April 2, 2013 at 7:14 am

I have backlinks from a website whose rank is 99.7 or something like that but on ahrefs its shows that the ahref rank of the backlink is 0.001, what exactly does that mean? could someone please explain I am new to backlinking. It was a blog comment backlin if thats any help?

Comments
Tony says
April 2, 2013 at 11:20 pm
Hello Tulisa. Where did you get the 99.7 rank of a backlink? Usually, a backlink and its weight can be measured by Page Rank, or by checking citation flow: http://stream-seo.com/real-time-page-rank/

Yes, blog comments do help, but you need a lot of them and to mix them with other backlinks. One comment won't really do anything to your blog.

Comments
Chetan Gupta says
March 22, 2013 at 2:08 pm
Thanks for this.
I liked it. I want to know some thing that Can i make comment on different Books of same blog for generating many backlinks to increase ranking. Will this trick will incease my ranking in google?

Comments
Tony says
March 22, 2013 at 2:12 pm

Comments are just a part of it. You need to distribute your backlinks within different strategies. Commenting on different Books is good, but it's better to have backlinks from different domains, not only one.

Comments
Mohi Uddin says
March 20, 2013 at 7:32 pm
Sir, I am a new blogger of Bangladesh, Please tell me How many back link needs for google page rank?

Comments
Tony says
March 22, 2013 at 9:50 am

For Page Rank, the best is to get Quality backlinks from high PR sites (e.g. PR2-PR7). It's difficult to know how many of them do you need, however, you can check that with Market Samurai in real time.

Comments
Tim Potter says
March 8, 2013 at 2:34 am
So much is made of back links. These really helps clear that up. It is really hard to know. I know keywords and site name propel you to high search engine result. Also it seems page rank is not as important as the other two for you to get good search results. Thanks for sharing

Comments
Tony says
March 11, 2013 at 11:14 am
Yeah, but it's also some kind of hit or miss. I'll update my experiment today and you'll find interesting results for week 4 ☐

Comments
Tim Potter says
March 11, 2013 at 11:19 am
I look forward to seeing the results. It will makes some interesting reading for sure. Again Thanks.

Comments
Tony says

March 11, 2013 at 11:42 pm
It's been updated now with Week 3 and Week 4 of the experiment! Getting good results on one Book and bad results on the other. Time to work on it!

Comments
Carl says
March 7, 2013 at 12:12 am
I Find this very Interesting, Not only did you provide me with useful information on how to build traffic, but you also show me out to track the process which will benefit you and other websites your up against. Are they any other products out their for tracking your website?, because im really new all of this.

Comments
Tony says
March 11, 2013 at 11:13 am
I normally use Market Samurai for tracking keywords and that stuff. But there are a lot of companies like SEO MOZ which could work for you.

Comments
Jennifer cunningham says
February 26, 2013 at 12:54 am
This is a good comparative study. I was reading in another forum and the person said their Books were usually 1,000-1,700 words and the site increased to page one after just 8 months and irregular Book posting. I'm beginning to think it's a combination of quality

content somewhere between 700-1,000 words and some quality backlinks. Backlinks take so much time to promote. I could use that time to write content. I'm hoping your experiment works and backlinks are not that important. I will google plus 1 this Book for your comparison. Come by funjobsforseniors.com.

Comments
Tony says
February 26, 2013 at 10:33 am
Hello Jennifer.
Yes. Long posts also tend to rank higher with time. From 1,000 words to more than 2,000. Backlinks take time to promote, but if you find a balance

between writing and promoting, you'll end up doing better with time.

Thank you!

Comments
Taswir Haider says
February 25, 2013 at 12:27 pm
Informative post ! Though there is no such critrea in order to determine number of backlinks for achieving top slots in search engine; still relevancy is the key of the success for ranking of keywords.

Comments
Tony says
February 25, 2013 at 12:39 pm

Thank you.
I have a nice update for this week. Going up all the way faster than I thought.

Comments
Arun says
February 19, 2013 at 8:28 am
Interesting n simple approach…Anyone can follow. Yes, it's definitely not about buying thousands of backlinks or using softwares and black-hat techniques, anymore. Miss the good ol days.lol. Thanks for sharing.

Comments
Tony says
February 20, 2013 at 10:44 am

I don't miss them. Because now people that work hard, will get their rankings high instead of people buying cheap links everywhere with low quality content.

Google isn't done yet. They still have a lot to improve before that happens.

Comments

Jason Hill says

February 18, 2013 at 5:50 pm

Hello Servando, great post! I am curious however, I hear a lot of people talking about backlinks and particularly "quality" backlinks. I am fairly new to blogging and more so to SEO and backlinks. How do you determine quality backlinks? How do you go about obtaining quality backlinks? I recently came across an Book that discussed backlinks and the method they talk about is creating several microblogs such as on blogger, wordpress.com, weebly etc. They then link those blogs to their main blog. They drive a lot of links to the microblogs which supposedly filters through to their money site. Is this a good strategy? What reccomendations can you make to someone new like me for building backlinks?

Comments
Tony says
February 18, 2013 at 6:02 pm
Hello Jason.

That's why I started this experiment. I want to know which kind of backlinks, how many of them, and what happens when sing only quality backlinks and then generic backlinks.

Ideally, a quality backlink would appear on another website related to your niche and have a good post that links (1-4 links maximum) to your original website. Those can be Guest Posts, Web 2.0 sites like WordPress, Hubpages, Ezine Books, etc.

But then again, it's just an experiment, and while I could make it to the top, I could also fail or just find a wall after reaching a rank. I don't know, and that's the beauty of this.

However, the good thing is that the last weeks my Books have been moving and this latest week has been the best now, even appearing at the first page of Yahoo results on one keyword.

If this backlinking strategy works, I'll definitely write guide for all my readers. As you can see on the SEOMOZ case study shared, they didn't use Tier 2 backlinks. But I'll test that too in the next 2 weeks.

Comments
Jason Hill says

February 18, 2013 at 6:06 pm
Awesome, look forward to following your progress Servando!

Comments
Avadhut says
February 21, 2013 at 12:27 am
Thanks Jason for asking this question. This is exactly I wanted to know.

Servando, please write a detailed guide on your strategy, so that we can learn this ☐

All the best, we want you to succeed.

Regards,
Avadhut

Comments
Tony says
February 21, 2013 at 7:14 pm
Hello Avadhut. I'll write a detailed guide if it works, of course.
If not, well..I'll write about it too.

Comments
Evan says
February 14, 2013 at 8:56 am
Thanks for the post! Indeed I asked this question so many times to myself. And after appeared in the middle of nowhere… Your post makes a complete sense to me.

Comments

Fahad Zahid says
February 14, 2013 at 4:36 am
Seems Innovative Idea Servando! Though there is no such critrea in order to determine number of backlinks for acheving top slots in search engine; still relevency is the key of the success for ranking of keywords.

Comments
Tony says
February 14, 2013 at 2:49 pm
Sure Fahad. and that's what I'm testing today.

Comments

anthony pietersen says
February 13, 2013 at 12:15 am
Good day Servando agreed with your post the links which show more relevancy,near to theme and appear natural to search engine have large life then dumb ones ….Keep updating

Comments
Tony says
February 13, 2013 at 12:33 am
I'll certainly do, Anthony. I've got a plan for my back-linking strategy that might work, and it's low risk (for this blog, at least).

But again, I won't focus on quantity here, and no black hat techniques.

Comments
Ana Hoffman says
February 12, 2013 at 10:04 pm
This looks interesting, Servando; keep us updated.

Comments
Tony says

Comments
Minhaj says
October 20, 2014 at 3:18 am

Well.. Remarkable Book and useful for us and we are waiting your another useful information about SEO and please provide about info Do follow link ..Absolutely100%WebsiteRanker!!!

Please!!
Leave a Comments & Reviews!!
Your email address will not be published.
copyright © 2016

www.ingramcontent.com/pod-product-compliance
Lightning Source LLC
Chambersburg PA
CBHW052243220526
45471CB00001B/175